FLOWERING PLANTS OF TAMIL NADU
MANGROVE ASSOCIATES

FLOWERING PLANTS OF TAMIL NADU
MANGROVE ASSOCIATES

D. Subramanian

Professor of Botany (Retd.),
Annamalai University,
Annamalai Nagar – 608 002,

Tamil Nadu.

Chennai　　　　　Trichy　　　　　New Delhi

MJP
PUBLISHERS

ISBN 978-93-88694-05-6 **MJP Publishers**

All rights reserved No. 44, Nallathambi Street,
Printed and bound in India Triplicane, Chennai 600 005

MJP 737 © Publishers, 2020

Publisher: C. Janarthanan

Project Editor: C. Ambica

PREFACE

In this book, the author has described a large number of species of mangrove forests and mangrove associates and 12 species of Marine Angiosperms with illustrations for most of them. Further cytotaxonomical and cytogenetical studies in most of the species of mangroves and the associates and 7 species of Marine Angiosperms of Tamil nadu have been added. For the first time the author has studied and published the cytological studies of them in cytologia and seaweed research and utilization journals and added these chapters in this book. A large number of mangrove associates, so far not reported, have been described in this book. The importance of these plants and further works necessary in them are also added.

This book is a part of author's extensive study of "Flowering Plants of Tamil Nadu - Mangrove Associates". It is hoped that this book will be useful to the coming generations of teachers, research workers and students of higher studies to have some more informations particularly about the mangrove vegetations and Marine Angiosperms of Tamil Nadu. As a cytologist, the author has revealed his experiences in his life time, about this very interesting and peculiar groups of plants and the author himself has drawn various plants and their parts in this book.

ACKNOWLEDGEMENTS

The author expresses his sincere thanks to Dr. A.L. Chidambaram, Dr. D. Thanavel and some of his Research Scholars, for having helped him in the collection of plants and cytological studies of them.

The author is also thankful to Dr. M. Palanisamy, Botanical Survey of India, Coimbatore, Tamil Nadu for helping him in getting references from the library of botanical survey of India. It is because of him, he could purchase the book "Seagrasses of coromandel coast, India by Ramamurthy et al., (1992)". This book is very much useful for understanding the earlier studies on this group of plants. In this book, there are illustrations of all the 12 sea grasses of Tamil Nadu coasts with diagrams of all the parts of each plant. This is very useful to identify all the 12 species of sea-grasses of Tamil Nadu. The author is also thankful to his friend, Mr. Jeeva and his son Mr. S. Shanmugam and Dr. Gopalakrishnan, Professor of Chemistry, Annamalai University for helping him in the collection of mangrove plants and sea-grasses from various places of Tamil Nadu.

The author likes to express his sincere thanks to Mr. B. Sivakumar, his research scholar, Department of Botany, Annamalai University, for his technical help in the cytological studies of Marine Angiosperms.

The author is thankful to Dr. Sundaramurthy, Professor and Head, and Dr. P. Ranganathan, Associate Professor, Department of Botany (DDE) Wing, for having helped him in studying the herbarium specimens. The help rendered by Dr. Kumarasamy and the non-teaching staff of this Department is gratefully acknowledged.

CONTENTS

PART – I

MANGROVE FLORA

INTRODUCTION

Mangrove formations are common in the tropics and subtropics but only a very few species are found in the temperate zones of the south and north hemispheres. Mangrove communities develop in the intertidal and subtidal regions but more so between arid-tidal level and extreme high water mark. The mangroves exist under very hostile and inhospitable conditions as they have to encounter extreme salinities, tides, of waves, wind velocities and temperatures. To overcome these adverse conditions, the mangroves are well adapted in their form and function. The mangroves possess halophytic properties such as thick cuticle, large mucilage cells, etc. in leaves, buttresses and knee, cable and stilt form roots. Further, vertical pneumatophores enable roots to get required supply of oxygen (Kannan, 1990).

Indian mangroves are distributed in about 6,740 sq.km. area (Krishnamurthy et al., 1987), constituting 7% of the world mangroves and 8% of the total Indian coastline. The Indian mangroves are one of the major forests of the Southeast Asia. More than 50 species of mangrove plants occur here, making the flora rich but with uneven distribution. There are three different types of mangroves in India viz. deltaic, backwater-estuarine and insular categories. The deltaic mangroves occur along the east coast (Bay of Bengal) where the mighty rivers (the Ganges, Mahanadi, Godavari, Krishna and Kaveri) make the deltas. The backwater-estuarine type of mangroves exist in the west coast (Arabian sea) which is characterized by typical funnel-shaped estuaries of major rivers (the Indus, Narmada and Tapti) or their backwaters, creeks and neritic inlets. The insular mangroves are present in the Andaman & Nicobar Islands where many tidal estuaries, "small rivers, neritic islets and lagoons support a rich mangrove flora (Gopal and Krishnamurthy, 1993). Mangrove forests of the Gangetic deltaic area and those of the Andaman & Nicobar Islands are the largest, constituting 75% of the total Indian mangroves. The remaining 25% is constituted by the other east and west coast mangrove forests.

Importance of Mangroves

Kannan and his Coworkers (1999) have explained the importance of Mangroves. Mangrove forests are useful to mankind in various ways as they yield both direct and indirect products of renewable nature. In addition, the mangrove environment provides living space for more than 200 species of epiphytes, invertebrates and fish. Many commercially important marine animals such as shellfish and finfish derive their food from the mangroves through detritus. Mangrove detritus not only supports the animal species of that area but also supports other animal populations of the adjoining areas after transportation from the mangrove area by currents (Krishnamurthy, 1975). Further, the mangrove environment serves as a very good nursery ground where the larvae of economically important animals such as prawns, crabs and fishes and still others take shelter and obtain food for their growth.

Mangrove environment also serves as the potential area for brackish water aquaculture of prawns, crabs and fishes as it happens to be Nature's own aquaculture system (Ong, 1982) which is more stable and less susceptible to diseases and epidemics. Small scale fisheries in mangrove waters produce nearly one million tones of fin-fishes, mollusks, crabs and shrimps annually, which is equivalent to about 1.1% of the world fishery catch (Kapetsky, 1985) and about one million people world over are dependent on mangrove-associated fisheries. The density of human population dependent on mangroves has been estimated as 5.6 persons Per km2 (FAO, 1988).

With regard to products, a few species of the back- mangrove community, for example, species of *Heritiera* and *Xylocarpus*, produce high-quality timber but their scarcity and difficulty of access have made them unattractive for harvesting. Exploitation of mangrove forest for timber is limited to only well managed forests like Gangetic Sundarbans, Andaman-Nicobar and deltas of Mahanadi, Godavari and Krishna. Mangrove trees are used variously for house construction, furniture, telegraph poles and other household articles.

Indian mangrove trees have 35% tannin in their bark which is higher than that of other countries (Roy Choudury, 1964). Good quality tannin can be extracted from the barks of *Rhizophora mucronata, Bruguiera gymnorrhiza* and *Ceriops tagal*. However, it should be noted that the mangrove tannin which was once the main source to the tannin industries of the country has been now replaced by synthetic tannin. The mangrove bark due to its tannin content is also frequently used by Indian fisherman to dye fishing nets to enhance their durability.

Honey collection from the mangrove forest is a promising business in India. It has been estimated that the Sundarban mangrove forest alone produces about 111 tones of honey actually (CIFR1, 1973). There are approximately 2000 people engaged in this trade. Honey collected from the *Cynometra ramiflora* and *Aegialitis rotundifolia* has a good market value and is in great demand. However, honey collected from other species like *Ceriops sp.* and *Excoecaria agallocha* is not highly valued (Blasco, 1975). Wax is another by- product from the mangrove forests and it is exploited by the government agencies. From *Nypa*, alcohol production from the sap of the inflorescence is about 15,000 litres\ha\yr while the sugar production is about 20 tones\ha\yr.

Only a very few mangrove plants are directly edible. *Bruguiera* and *Avicennia* seeds, if cooked and boiled, can be eaten as famine food. While leaves of *Nypa* palm are used for thatching roofs, *Suaeda* and *Acrostichum* leaves are used as green vegetables. Salads are made from the leaves of *Sonneratia* and *Sesuvium*. Radicles of *Rhizophora, Bruguiera, Sonneratia* and *Ceriops* species, seeds of *Heritiera littoralis* and tender leaves of *Acrostichum aureum* and *A. speciosum* are consumed as food. The fruits of *Nypa, Avicennia. Sonneratia* and *Heritiera* are also eaten. Some epiphytes like *Hydnophytum* provide fruits for birds.

Barks and roots of *Aegiceras corniculatum* and *Derris heterophylla* are used as fish poison (saponin). Barks of *Derris elliptica, D.scandens* and *Aegiceras corniculatum* and milky latex of *Excoecaria agallocha* are used as fish poison, particularly by the tribals in the Andaman & Nicobar Islands (Dagar, 1982, 1987).

Different mangrove species are used by the local people for medicinal purposes. Some of the plants are used as astringents, antidiarrhoeal, purgative and anthelmintic agents and also as antidotes for venomous bites. Still others are used to treat skin diseases, leprosy, blood pressure, diabetes, malaria etc.

The Mangrove forests of India are natural forests found from time immemorial. After the scientific developments in 20th century, people began to realize the importance of mangrove forests. But, at the same time increase of human populations, deforestation, use of mangrove woods as fuels and natural disasters, cause great disturbances leading to total destruction of mangrove forests. During the Tsunami attack in December 26, 2004, most of the seashore villages escaped near Killai and Pichavaram because of these forests. Now, protection measures are tightened in these areas and perhaps, these same protections of mangroves will be extended throughout the world.

Dr. A.Abdul Rahman (1990), Professor of Zoology at Poondi Shri Pushpam College, Tanjore District has made an extensive investigation on mangrove vegetations of Tamilnadu, particularly Muthupettai estuaries (1987 & 1989) of Tanjore District. He has described many floras and faunas of mangrove forest of Tamilnadu. He has written a book in Tamil namely "The Fertility of Mangrove Forests".

Dr.L.Kannan and R. Kannan (1999), Department of Marine Biology, Annamalai University, has written a book "How to know the Mangroves?" and enumerated some of the species of mangrove plants from India.

The author published a research paper in "Cytologia" journal in 1988 entitled "Cytological studies of some mangrove flora of Tamilnadu".

Of the 16 species of mangrove plants cytologically studied, first records of chromosome numbers have been made in 9 species by him.

Mathew (1982) has described a few plants of Mangroves in his book "The flora of the Tamilnadu Karnatic". Apart from the above mentioned latest works on mangrove forests, no other informations are available so far. Therefore, the present contribution may be more useful to the Teachers, Scientists, and animal and plant breeders and to those interested in the study of mangrove forests.

In this book, the morphological features of mangrove plants and the taxonomical characters of them are provided. Key preparations for various species studied are also given.

India

The mangrove forest area of India is 6740 sq.km. It has occupied only 7% area of world's total mangrove forests (World's total mangrove forest area is 99,300 sq.km).

	State / Union Pradesh	Area (Sq.km)
1.	Andaman Nicobar	1190
2.	Bengal	4200
3.	Orissa	150
4.	Andhra	200
5.	Tamilnadu	150
6.	Karnataka	60
7.	Goa	200
8.	Maharastra	330
9.	Gujarat	260
	Total	6740

One of the most important World's mangrove forest is present in Bengal (Blasco, 1977) called Sundarban forests with an area of 5900 sq.km. Of these, the mangrove forest area is approximately 1190 sq.km. and it has 29 plant family of 41 genera and 60 species. Among these, 25 species are found in Bay of Bengal and 8 species only found in Arabian sea. The plant distribution depends on the important factors like rainfall, temperature and salinity.

Tamil Nadu

Various economic values are available from Tamilnadu mangroves. According to Venkatesan (1966) Tamilnadu had consisted of 148.97 sq.km area of mangrove forest, once upon a time and believed that the mangrove forest was distributed fully on the Cauvery delta coastal region from Cuddalore to Kodiakarai and nearby region of Sathiram. As such, the mangrove forest was present on Tuticorin and Pamban region also. The brief explanation of mangrove forest in Tamilnadu is given below.

In 1882 published Cuddalore district map was the first evidence for the identification of Tamilnadu mangroves. This map has shown the mangroves in the northern end of Cauvery delta and Killai, Pichavaram. However, this evidence needed various schemes of forest conservation facility. Thirumalai Raj's research articles published in 1959 described the various details of scientific informations. After 1966, Venkatesan described the names of mangrove plants and their ecological factors. Further many National and International researchers were interested to study these mangroves, noteworthy among them are Raja Gopalan (1952), Mager-Homiegie (1973) and Karadini's (1973). Ganapathy Thanigai Mani of Pondicherry French Institute also described the archaeology of mangrove plants in Tamilnadu briefly.

In 1963 Chithu had valued 26500 hectare area alone occupied by mangroves. In 1966 Venkatesan proposed that 14,897 hectare mangrove land area existed. But in 1959 Thirumalai Raj considered that the mangroves spread in very large area. According to him, nearly 45,000 hectare area was occupied by the mangrove forest and among them only 1000 hectare area is found as dense forests. The remaining region was fully occupied by stunted bushes and shrubs only. The mangrove forests' actual area of Tamilnadu has not been known so far. It may measure approximately 500 to 1000 hectare. But, today the mangrove forests of Tamilnadu are considered to be in totally destroying condition. Flasco (1975) considered that human disturbances may be more important reason for the destruction of mangrove forests.

The Vellar river originated from the Servarayan hill stations of Salem district runs through South Arcot district and finally reached the sea in Parankipettai. Also the river of Ponni emerging form Kudagu mountain enriching Karnataka state and flowing through Tamilnadu, finally emerges as a branch namely Kollidam and engulf into the sea in Tanjore-South Arcot districts. In between the estuaries the small river of Uppanar and Puckinkam canal mingled here. This complex formed for the Killai - Pichavaram mangroves. This region is located on 17°29' North latitude and 79°46' East longitude.

The estuarine region of Muthupettai located on 10°46' N latitude and 79°51' E longitude. The Vennar river is important in East Tanjore district. It has many branches. Among them the important branches are Pamani river, Korai river. These two rivers are not directly mixed in the sea. Finally they are mixed together forming the Muthupettai lagoon. The branch of Korai river diverted into various branches namely Mulli river, Marakka korai river, Ponnukkundan river, Valavan river, killai thanki river, Harichandra river, Paluvan river, Nallaru river and finally they mixed in the sea at Vetharanyam Taluk. The later four branches are responsible for the facility of Thalaignairu mangroves.

In Tamilnadu, the mangrove forests are present in the river bank of Ariyankuppam and Kalveli Tank edges of Pondicherry. They are also found at Rameswaram and Islands of Gulf of Mannar as small patches. These are all remnants of once large establishments.

At Muthupettai there is increase of these forests approximately to 400 hectares. There is also increase of mangrove forests at Thambikkottai areas. Like this, at the estuaries of Pazhiar-Nasuviniyar areas, there are mangrove forests. There is also mangrove forest at Adirampattinam sea-shores.

Rainfall, hurricanes in raising the heavy floods of Cauvery, seawater infusion, deposition of salt etc. are the important factors of establishing mangrove plants.

CYTOLOGICAL STUDIES

The cytological studies in Mangroves of India by the author (1988) is the first attempt in India and possibly all over the world. Only there are fragmented reports in one or two species like Sesuvium portulacastrum, Rhizophora mucronata, Acanthus ilicifolius, Salvadora persica, Pavetta indica and Ipomoea biloba. But, excepting Rhizophora mucronata, all the other above said plants can live in soil with little amount of salt or even without salt. Therefore, in true mangroves, cytological studies have been made only in Rhizophora mucronata (Fedorov, 1974). As far as the author is aware, no cytological studies have been made in other mangrove plants of the world so far. Therefore, the present study on mangroves is more important, as the chromosomes of plants are the seats of hereditary characters and by means of that, it is possible to understand the interrelationships and evolutionary trends of them.

Cytological studies are difficult in tree plants. Further, it is more difficult in mangrove plants, as they deposit a large amount of salts, alkaloids and various other chemical substances. The roots are leathery. Therefore, the normal acetocarmine or aceto orcein methods, which are successful for most of the flowering plants are not applicable to the root — tip squashes of mangrove plants. Therefore, the author tried to find out a suitable staining method by means of continuous laboratory experiments. It took 6 to 8 months. But even now, it is not possible to take microphotographs, as the chromosomes are faintly seen.

All the seedlings of the taxa studied were collected from mangrove forests of Killai and Pichavaram of Tamilnadu and grown in the Botanic Garden, Annamalai University for 7 to 15 days to get the root tips, excepting *Rhizophora candelaria, R. mucronata* which were grown for 6 months before commencing the cytological investigations in order to avoid to some extent the presence of excess of alkaloids and latex secretions in the root tips.

Young and healthy root tips were pretreated in 0.02% hydroxyquinoline kept at 4°C for 3 hours. After a thorough wash, they were fixed in 1:3 acetic ethanol for 12 hours. Then, they were washed and softened in conc. HCl for 5 to 10 minutes till they were sufficiently softened. After a thorough wash, they were kept in Petri dishes containing 1% aceto orcein for an hour. With 1 or 2 drops of aceto orcein and a new coverslip, one or two root tips were squashed per slide and sealed. Camera lucida diagrams were drawn under oil immersion lens under a magnification of x2500.

In *Sesuvium portulacastrum,* globular leaf type has 36 chromosomes, whereas the lanceolate leaf type possesses 40 chromosomes. The prophasic chromosomes are found dispersed within a circular configuration. Besides, there are anaphasic bridges and precocious movements of chromosomes rarely observed in this species In *Rhizophora candelaria* and *R.mucronata,* somatic chromosome number is 2n=36. In the former species, during metaphase, there are rarely precocious movements of chromosomes. Besides, there are frequent amitotic divisions resulting into the formation of nuclear fission and micronuclei formation. In *Ceriops roxbughiana, Avicennia marina* and *Aegiceros corniculatus* there are 2n=36 chromosomes. In *Ceriops roxbughiana,* there are rarely anaphasic laggards. In *Bruguiera conjugata* there are 2n=38 chromosomes. There are 2n = 18 chromosomes in *Suaeda maritima, S.monica* and *Arthrocnemum fruticosum*. In *Acanthus ilicifolius, Salvadora persica, Pavetta indica* and *Ipomoea biloba*, the present report of chromosome number has been confirmed by the earlier records (Table 1.).

The following categorization of chromosomes has been made, with a view to describe the karyotype and to represent the same by karyotype formulae. This facilitates the understanding of interspecific and intergeneric relationships of the taxa analysed.

Type A : Longer chromosome (4.0 mm to 5.0 mm) with submedian primary and subterminal secondary centromeres.

Type B : Longer chromosome (4.0 mm to 5.0 mm) with submedian centromere.

Type C : Longer chromosome (4.0 mm to 5.0 mm) with median centromere.

Type D : Long chromosome (3.0 mm to 3.9 mm) with median primary and subterminal secondary centromeres.

Type E : Long chromosome (3.0 mm to 3.9 mm) with submedian centromere.

Type F : Long chromosome (3.0 mm to 3.9 mm) with median centromere.

Type G : Medium sized chromosome (2.0mm to 2.9mm) with median primary and subterminal secondary centromeres.

Table 1 Chromosome Numbers of Mangrove Plants in Tamilnadu

S.No.	Species studied	2n chromosome number	Previous report of chromosome numbers, authors and years
1.	*Sesuvium portulacastrum* globular elongate leaf type, Aizoaceae	36	36, Sharma and Bhattacharrya 1956 48, Raghavan and Srinivasan 1940
2.	*Sesuvium portulacastrum* Lanceolate leaf type, Aizoaceae	40*	-
3.	*Rhizophora candelaria* Rhizophoraceae	36	-
4.	*Rhizophora mucronata* Rhizophoraceae	36*	36, Patil 1958
5.	*Ceriops roxbughiana* Rhizophoraceae	36*	-
6.	*Bruguiera conjugata* Rhizophoraceae	38*	-
7.	*Avicennia marina* Verbenaceae	36*	-
8.	*Suaeda maritima* Chenopodiaceae	18*	-
9.	*S. monoica* Chenopodiaceae	18*	-

S.No.	Species studied	2n chromosome number	Previous report of chromosome numbers, authors and years
10.	*Arthrocnemum fruticosum* Chenopodiaceae	18*	-
11.	*Acanthus ilicifolius* Acanthaceae	44	44, 48 Narayanan 1951a, b
12.	*Salvadora persica* Salvadoraceae	24	24, Koul and Singh 1962
13.	*Aegiceros corniculatus* Myrsicaceae	36*	-
14.	*Pavetta indica* Rubiaceae	22	22, Homeyer 1932, Fagerlind, 1937
15.	*Ipomoea biloba* Convolvulaceae	30	30, King and Barnford 1937, Sugiura 138, Sharma and Datta 1958
16.	*Ilysanthus tenuifolia Urb* Scrophulariaceae	16*	-

*First record of chromosome numbers in the respective species.

Type H: Medium sized chromosome (2.0 mm to 2.9 mm) with submedian centromere.

Type I: Medium sized chromosome (2.0 mm to 2.9 mm) with median centromere.

Type J: Medium sized chromosome (2.0 mm to 2.9 mm) with subterminal centromere.

Type K: Short chromosome (1.0 mm to 1.9 mm) with median primary and subterminal secondary centromere.

Type L: Short chromosome (1.0 mm to 1.9 mm) with submedian centromere.

Type M: Short chromosome (1.0 mm to 1.9 mm) with median centromere.

Type N: Short chromosome (1.0 mm to 1.9 mm) with subterminal centromere.

Type O: Shorter chromosome (less than 1.0 mm) with subterminal centromere.

1.   *Sesuvium portulacastrum*, globular-elongate leaf type

   Total chromosome length = 50.6 mm

   Range = 2.0 mm to 1.2 mm

   2n = 36 = G2+K2+L4+M14+N14

   The present report of chromosome number is in confirmation with the earlier report of chromosome number (36) by Sharma and Bhattacharrya (1956) but differs from that (48) of Raghavan and Srinivasan (1940). There are a few median sized and mostly short chromosomes.

2.   *Sesuvium portulacastrum* (lanceolate leaf type)

   Total chromosome length = 92.4 mm

   Range = 4.0 mm to 1.2 mm

   2n = 40 = A2+E2+H2+I18+N14

   The karyotype study shows that there is a gradation of chromosomes from longer to short size.

3.   *Rhizophora candelaria*

   Total chromosome length = 43.6 mm

   Range = 1.6 mm to 1.0 mm

   2n = 36 = K2+L4+M4+N16

   There is no earlier record of chromosome number for this species. All the chromosomes observed are of short type.

4.   *R. mucronata*

Total chromosome length = 39.4 mm

Range = 1.5 mm to 0.8 mm

2n = 36 = K2+L10+M4+N8+O12

The present report of chromosome number is in conformation with the earlier report (2n = 36) by Patil (1958). The chromosomes studied are of short and shorter types.

5.   *Ceriops roxburghiana*

Total chromosome length=53.6 mm

Range = 2.0 mm to 1.2 mm

2n = 36 = G4+M16+L4+N12

No earlier report of chromosome number or karyotypic study in this species. There are a few median sized chromosomes and others are short sized.

6.   *Bruguiera conjugata*

Total chromosome length = 56.0 mm

Range = 2.8 mm to 1.2 mm

2n = 38=G4+I12+N22

No earlier report of number or karyotypic study in this species. The chromosomes are medium sized and short.

7.   *Avicennia marina*

Total chromosome length = 57.2 mm

Range = 2.0 mm to 1.4 mm

2n = 36 = G2+M18+L6+N10

No earlier report of chromosome number or karyotypic study in this species. There are a few median sized chromosomes and others are short.

8.   *Suaeda maritima*

Total chromosome length=57.2 mm

Range = 2.0 mm to 1.2 mm

2n= 18 = G2+M6+N10

No earlier report of chromosome number or karyotypic study in this species. There are two medium sized chromosomes and the rest are short.

9.   *Suaeda monaica*

Total chromosome length = 22.0 mm

Range = 1.6 mm to 1.0 mm

2n= 18 = L4+M8+N6

No earlier report of chromosome number or karyotypic study in this species. All the chromosomes are short

10.   *Arthrocnemum fruticosum*

Total chromosome length=67.2 mm

Range = 5.0 mm to 2.5 mm

2n = 18 = A2+B2+C6+E2+F2+J2

No earlier report of chromosome number or karyotypic study in this species. There are a graded series of longer, long and medium sized chromosomes.

11.   *Acanthus ilicifolius*

Total chromosome length=60.4 mm

Range = 2.0 mm to 1.0 mm

2n=44=G2+M10+L16+N16

The present report of chromosome number has been confirmed by the earlier record of chromosome number (2n=44) by Narayanan (1951a,b) in this species. Only two chromosomes are medium sized and the rest are short.

12.   *Salvadora persica*

Total chromosome length=23.6 mm

Range = 2.0 mm to 0.8 mm

2n = 24 = L6+M10+O8

The present report of chromosome number has been confirmed by the earlier report of chromosome number (2n = 24) in this species (Table 1). The chromosomes are of medium and short size.

13. *Aegiceros corniculatus*

Total chromosome length = 40.2 mm

Range =1.5 mm to 0.8 mm

2n = 26 = K2+L14+M10+N4+O6

No earlier report of chromosome number or karyotypic study in this species. Only 6 chromosomes are shorter and the others are short.

14. *Pavetta indica*

Total chromosome length = 26.0 mm

Range = 1.3 mm to 1.0 mm

2n = 22= L10+M4+N8

The present report of chromosome number has been confirmed by the earlier report (2n = 22) in this species by Homeyer (1932) and Fagerlind (1937). All the chromosomes studied are short.

15. *Ipomoea biloba*

Total chromosome length= 79.6 mm

Range = 3.8 mm to 2.0 mm

2n = 30= D2+E2+F6+H10+I2+J8

The present report of chromosome number in this species is in confirmation with earlier record of chromosome numbers (Table 1). The chromosomes are medium sized and short.

16. *Ilysanthes tenuifolia*

Total chromosome length= 18.6 mm

Range = 1.4 mm to 0.9 mm

2n=16= L6+M6+03

No earlier report of chromosome number in this species. The chromosomes are short and shorter.

16 taxa coming under 10 families of mangrove vegetation from east coast areas of Tamilnadu have been cytologically, morphologically investigated, of which first record of chromosome numbers has been made in 10 species. Deviant reports as against (Table 1) the earlier records have been made in *Sesuvium portulacastrum* (2n = 48, Raghavan and Srinivasan, 1940) and in *Acanthus ilicifolius* (2n = 48) (Narayanan, 1951b). In other species studied the

present reports of chromosome numbers coincide with those of the earlier investigations.

There is a range of somatic chromosome numbers among the species studied from 2n = 16 to 2n = 44. But, 2n= 18 and 36 are of more common occurrence. The number 2n = 18 occurs in 3 species and the number 2n = 36 in 6 species. There are 2 distinct lines of evolution among the species studied, one represented by 2n = 18 chromosomes and the other by 2n = 36 chromosomes. This is evidenced by the study of frequency of diploid chromosome numbers in various species of mangrove flora studied.

The basic chromosome number of the family may be n = 18, as 2n = 36 occurring in maximum number of species of mangrove vegetation. From n = 18, the basic numbers n = 8, 9, 11, 12 and 15 should have come about by reduction in chromosome numbers such as visualized in *Crepis* (Babcock 1947). Tobgy (1943) has shown that the 3 paired *C. fuliginosa* was derived from the 4 paired *C. neglecta* through a system of reciprocal translocation. Some such processes might have been in operation in the derivation of lower basic numbers from n = 18. In the evolution of species having more basic numbers from n = 18, a process of duplication of the chromosomes and reorganization of broken chromosomal portions take place. It is concluded that both aneuploidy and euploidy played an important role in the origin and evolution of the species studied.

The chromosomes are generally short and shorter, an advanced character. Excepting *Arthrocnemum fruticosum*, the number of submedianly and subterminally constricted chromosomes (including the secondary constricted chromosomes) are far exceeding the number of medianly constricted chromosomes in the respective species studied. This is another cytologically advanced character of the species included in the present investigation. These advanced cytological characteristics together with polyploidy may be the tools for the species to tolerate the heavy salt deposition of the Mangrove vegetation.

MORPHOLOGICAL STUDIES

The author has visited most of the areas of eastern coast of Tamil Nadu and collected the plants and studied the important morphological and cytological characters of both mangroves and mangrove associates. All the diagrams have been drawn by the author himselves for cytological and morphological studies. Some of the herbarium specimens of botanical survey of India at Coimbatore, Tamil Nadu have been used for identification and diagrams. Similarly, the herbarium specimens of mangroves at Botany Departmnt, DDE Wing, Annamalai University have been used for diagrams and verifications.

FERNS: Achrostichum aureum and A Speciosum are the ferns living with Mangrove Plants (Plates 1 and 2).

Sonneratiaceae

Trees. Leaves decussate, coriaceous, Inflorescence 3 in a cluster or sometimes flower 1; flowers 4 to 6-merous, bisexual, regular, stamens many in several series, inserted on the rim of the calyx tube, filiform - subulate. Ovary sessile, 4 to 6 or more celled, ovules axile. Berry indehiscent, Seeds many.

This family represented by only one species.

Sonneratia apetala Bach-Ham. (Plate 7)

Tree, 7 to 9 meters, Leaves decussate, oblong or linear – lanceolate; Flowers 3 in a cluster, cream coloured, calyx lobes 4, persistent; Stamens many in several series, Petals absent; Ovary 4 to 6 celled, free from calyx, ovules many, axile placentation; Berry globose. Seeds many, curved, angular.

One of the mangroves in backwater canals in Cuddalore, Pichavaram, Killai forming a dense colony, common.

Combretaceae

Lumnitzera Willd. is one of the 6 genera coming under this family and it is a mangrove plant having only one species L.racemosa.

Lumnitzera racemosa, Willd (Plate 6)

Shrub, 4 to 6 meters in height; shining green; Leaves variable in shape; Inflorescence spike, bracteoles 2, adnate to the base of the calyx; Flowers pentamerous, calyx tube produced beyond ovary, lobes 5, petals 5, white, disc obsolete; Stamens 10 in 2 series, ovary one celled, oblong, ovules 2-5; Fruits ellipsoid, Woody.

Mangrove plant in backwater canals of Eastern and Western Sea-shores of India, rare, Also recorded in mangrove forests of various countries of the world.

Salvadoraceae

There are two genera, Azima and Salvadora L. Salvadora has but one species S.persica L.var. wightiana (Thwaites) Verde. This species, even though present in scrub jungles, is also growing well in Killai along with other mangroves. A plant is present from the past 30 years in Killai tourist center near Chidambaram. Therefore, this plant is described here as one of the members of mangroves of Killai and Pichavarm.

1. *Salvadora persica L.var. wightiana (ThwaitesVerde.* *(Plate 5)*

Tree with hanging branches, sometimes touching the ground on all sides, 8 to 12 meters in height, compact crown. Leaves elliptic - oblong, light green, shining. Flowers bisexual, axillary or terminal panicles. calyx 4 lobed. cupular, corolla cream coloured shortly tubular, lobes 4, recurved; Stamens epipetalous, 4, ovary unilocular, one seeded, plains scrub jungles and mangrove forests of Tamilnadu.

2. *Azima tetracantha. Lam. (Plate 4)*

It is one of the mangrove associates, present at Muthupet of Tanjore district and Kille and Pichavaram of Cuddalore district of Tamil Nadu.

Rhizophoraceae

There are 4 genera namely Rhizophora, Bruguiera, Ceriops and Kandelia. Of these, the former three genera are common and Kandelia is very rare and an endangered species fastly disappearing from our world.

1. *Bruguiera Lam.*

There are 2 species, B.cylindrica and B.gymnorrhiza and both of them are small trees. Pneumatophores are numerous with knee like knots then and there arising above the sandy mud. The hypocotyl of the fruit slightly curved and sometimes ribbed.

i) *B.cylindrica* (L.) Blume. (*Plate* 8)

Bushy small tree, 3 to 6 meters in height, leaves elliptic-oblanceolate, 5.5 -10 x 2-3 cm in size; Cymose inflorescence 3 flowered. calyx lobes 8 or 9, reflexed in fruit petals 9, greenish yellow, margins ciliate, bristles 2 or 3 per lobe, fruit viviparous hypocotyls to 8 or 9 x 0.7 to 0-9 cm, radicle dark green, along with Rhizophora and Ceriops on the shores of back water canals of Killai, Pichavaram, abundant.

ii) *B. gymnorrhiza* (L.) Savigny (B.Conjugata Merr.) (*Plate* 9)

Bushy small trees, 3 to 5 meters in height; Leaves lanceolate, 7.5 to 15 x 3.6 to 5 cm in size. Flowers 3.5 cm across, drooping, Calyx- lobes 9, leathery, petals 9, reddish outside, cream inside, outer margin fringed towards base with white silky hairs; Stamens 18, 9 longer and 9 shorter, basifixed, anthers oblong; Ovary superior, axile placentation, 4 loculed; one ovule in each locule, fruit with brown radicle and green and cylindrical hypocotyl.

2. *Ceriops* Arn

This is a large shrub resembling the species of Bruguiera particularly in general morphology of the plants and nature and appearance of flowers and fruits. However, this genus differs from Bruguiera in small size of plants, leaf shape and angular fruits.

1. *C. decandra (Griffith) Ding Hou.* (*Plate* 11)

A large shrub, 2 to 4 meters in height, forming rounded bushes; branches stiff, brittle; Leaves apically clustered, oblong - ovate, obscurely nerved, not punctuate; Inflorescence condensed cyme, capitate; 5 to 8 flowered; bracteoles 2 at the base of the flower, cupular, calyx lobes 5, ovate, petals 5, longitudinally folded, cream to greenish - yellow, margins pubescent, apex fringed, disc annular and lobed; Stamens 10, inserted at the lobes of disc; 5 longer and 5 shorter, basifixed, anthers with tailed tip; Ovary 3 celled, ovules 2 in each locule; Fruit ovoid-conical to 1 cm; Cotyledons 1 to 1 cm, hypocotyl angular, sulcate.

A common mangrove plant along with *Rhizophora species* and *Bruguiera cylindrica*. It is gregarious.

2. *Ceriops tagal (Plate 10)*

This species was collected from Pichavaram of Cuddalore district; Leaves opposite, broad with circular apex. In each node of leaves cluster of flowers or fruits present. The epicotyl smooth and cylindrical.

3. *Ceriops candolleana (Plate 10)*

This species is also available at Kille and Pichavaram of Cuddalore district. The leaf broad with narrow tip. At each node a pair of flowers or fruits present. The epicotyl grooved and cylindrical.

3. *Kandelia (Dc.) Wight & Arn.*

It is one of the many flowering plants facing extinction. Even in Botanical Survey of India, Southern circle at Coimbatore, a few (2 or 3) herbarium specimens alone are available and that too collected from Cuddalore and Pichavaram of Tamilnadu. But now, we could not see it from these areas for the past 20 years. There is but one species.

Kandelia candel (L.) Druce. (Plate 17)

(*K. rheedii* Wight & Arn.)

It is a large shrub or small trees. Leaves elliptic oblong, 9.5 to 11 x 2.5 to 5 cm in size; cymes 4 to 9 flowered; bracteoles 2-4, copular, flowers 3 cm across, calyx lobes 5, oblong, reflexed in fruits, Petals 6, oblong, 2 lobed, lobes multifid, sinus with a long bristle; Stamens many, free; Ovary 1-celled, ovules 6; Stigma 3-lobed; Fruit ovoid, narrow above broader in the middle and sharply narrowed into a sharp point. In this character and in leaf and flower morphology, Kandelia candel differs from Rhizophora mucronata and R.apiculata. In general appearance Kandelia candel looks like the species of Bruguiera and Ceriops. The author has seen the herbarium specimens of this species in Botanical survey of India at Coimbatore. M.S. Swaminathan and Sanjan Deshmukh (1995) have given the photographs of the portion and entire plants of this species as seen in Goa, India. The author tries to locate this species in Tamil Nadu.

4. *Rhizophora L.*

It is noteworthy at this point that pneumatophores, stilt roots and viviparous fruits — all these characters are associated with mangrove plants particularly in all 4 genera of Rhizophoraceae. Above all, these adaptations are more pronounced in Rhizophora. Another important point is that there are a large number of distinct populations among the 2 species, R. apicutata and R. mucronata. This may be due to frequent interbreeding among the various populations of the two species. Therefore, care should be taken in establishing new varieties and species in Rhizophora. This phenomenon is more pronounced among the species of Rhizophora in Killai and Pichavaram mangrove forests.

The two species are large trees in this genus with longer hypocotyl than the other mangrove species.

1. *Rhizophora apiculata.* **Blume. (Plates 13 and 14)**

(*R. candelaria* **Dc.**)

Large trees, 10 to 15 meters in height. Leaves thick green, lanceolate, 5 to 13 x 3.5 to 5.5 cm in size, strong midrib, pinnate venation, pointed tip; Cymes 2 flowered, peduncles much reduced, bracteole cupular, crenulate; Flowers greenish yellow, calyx lobes 4, lanceolate; Petals green outside, yellowish inside, oblong, leathery; Stamens 10 - 12, sessile; Fruit viviparous, hypocotyl large obconical. Stilt roots large and numerous forming impenetrable barriers. Abundant in Killai and Pichavaram mangroves.

2. *R. mucronata.* **Poiret. (Plates 12 and 13)**

Large trees, 8 to 15 meters in height; Leaves light green, oblong – elliptic, 7 to 15 x 4 to 7.5 cm in size; Cymes repeatedly branched; peduncle to 6 cm; bracteoles 2 lobed, lobes deltoid; acute; Flowers large, 2 cm across; petals lanceolate, densely pilose along margins, glabrous on the back; Sepal thick leathery, 4; petals 4; Stamens 8; Fruit ovoid smaller than that of R.apiculata. Ovary 2-celled; axile placentation, 2 ovules in each cell; Stilt roots lesser and lesser spread of crown than that of R.apiculata.

Swaminathan M.S. and Sanjay Deshmukh (1995) have provided a photo of a twig of *Rhizophora* hybrid (*R. apiculata* x *R. mucronata*) but it is one of the populations, different from one another from Tamil Nadu. (Plates 15 and 16)

Myrsinaceae

Aegiceras corniculatum Linn and A. majus (Plates 20 and 21) are present at kille and Pichavaram. They have clusters of curved fruits

Aizoaceae

Sesuvium Portulacastrum is a Succulent Plant at kille (plate).

Chenopodiaceae

Arthrocnemum indicum (Plate 27) and Suaeda monoica (Plate 28) and s.nudiflora (Plate 29) are common succulent herbs present among the mangrove vegetations through out Tamilnadu.

Avicenniaceae

Mangrove trees or shrubs, roots with numerous, erect pneumatophores; Leaves decussate. Only one genus under this family Avicennia.

Avicennia L.

Shrubs or trees, branchlets tetragonous, swollen at nodes often with galls; Leaves coriaceous; Cymes in pedunculate heads, in turn to tri–chotomous panicles. There are two species.

1. *A. marina* (Forsskal.) Vierh. (Plate 25)

Shrubs to 4 to 5 meters in height; Leaves obovate to oblanceolate; 6 to 10 x 2.5 to 4cm in size; corolla yellow; 4 petals; Stamens included, 4 stamens; basifixed; Ovary superior, axile placentation, one ovule in each locule. Drupe 1.5 x 1 cm in size.

2. *A. officinalis* L. (Plate 26)

Trees to 15 meters in height, with a hemispherical crown; Leaves broadly ovate to rounded, 4 to 9 x 4-6cm in size; peduncle to 10 - 12 cm; Corolla 5 petals, yellow, 1.2 cm across; Stamens 5, exerted; Drupe 2 x 1.5 cm in size along with A. marina.

Both these species are available in Killai and Pichavaram of Tamilnadu.

Acanthaceae

Acanthus L. is one of the 32 genera represented in Eastern Ghats, Western Ghats and Plains of Tamilnadu under this family. This genus has but one species, A.ilicifolius L. and it is a mangrove plant. During the last tsunami attack in South India in December 26, 2006, almost all the thickets of this species were destroyed in Killai and Pichavaram. It is collected now in backwater canals of Alapakkam village near Cuddalore.

Acanthus ilicifolius L. (Plate 23)

Gregarious thorny shrubs to 1 to 1.5 meters in height forming a colony or thicket into which, it is very difficult to enter; Leaves long, pinnatifid, thick coriaceous, lobes zig-zag, spine tipped, petiole short and stout with a pair of short based spines, spikes strobilate, terminal, peduncle 7 cm, basal bracts 2, bracteoles 2, calyx lobes 4, coriaceous, 2 + 2; Corolla brightly bluish purple, large, upper lip obsolete, lower lip shortly 3 - lobed; Stamens 4, exerted, anther walls longer and bearded, one cell sterile; Ovary axile placentation, superior, 4 ovuled, capsule almost globular, 0.4 x 3.5 cm; Seeds 4, orbicular.

Available in almost all South-east Asiatic countries and Australia.

Euphorbiaceae

In this family, there is one mangrove species namely Excoecaria agallocha L. The Tamil name of this species is "Thillai" and once upon a time there was a large forests mostly occupied by this species from Chidambaram to eastern sea-shore of Bay of Bengal. The supreme Goddess "Thillai kali" was conquering this forest and its surrounding region and so it was called "Thillai forest". Even now, in some of the lagoons and shores of backwater canals, this plant is present and it is a predominant species in Killai and Pichavaram mangroves even today. It is an important medicinal plant in ancient country medicine in Tamilnadu.

Excoecaria agallocha L. (Plate 30)

Large shrub or small tree sometimes; Leaves alternate, elliptic - lanceolate, 5 to 9 x 3 to 4.5 cm; Male spikes to 12 cm; Female spikes to 4 cm, Capsule pendulous with weak pedicel, 3 cm across, 3 chambered; Ovary trilocular with one ovule in each, axile placentation; 3 stamens in each male flower, no well defined petals; The stem and leaves on wounding give alkaloid latex which is harmful to our eyes and makes them blind if applied. But in lower concentrations this juice is a powerful medicine.

Poaceae

There are a few genera of grasses present usually in saline soil near seashores forming separate patches or prostrate mats.

1. *Coelachryum lagopoides (Barm. F.) Senar.*

(*Eleusine lagopoides* (Burm.f.) Merr. (Plate – 36)

Culms creeping, 10 cm long, glabrous at nodes; Blades oblong - lanceolate, plicate; Spikes dense, aggregated, globose, Spikelets secund, several fid, disarticulating between florets; Glumes lanceolate, subequal , strongly keeled, rounded at the back, stiff aristate, lemma aristate, 3-nerved, 3-keeled; appressed along keels at back keels, winged. Anthers 3, plains at salt soils, present at saline soil of Sea- shore areas of India.

2. *Aleropus Lagopoides (L.) Trincex. Thwaites (Plate – 35)*

Culms of 10 cm, creeping, blades lanceolate, subdistichous; Panicles a globose head of densely clustered spikelets, pinkish, spikelets elliptic-oblong, laterally compressed, 4 to 8 flowered, Glumes unequal, lemmas oblong-ovate, mucronate; Anthers 2, Ovary ellipsoid.

Inhabit in Coastal areas of Eastern and Western Sea-shores of India forming prostrate mats.

3. *Portiresia coarctata (Oryza coarctata) (Plate – 34)*

It is present in large scale at mangrove vegetations of Calcutta. But the author has seen this plant in a swamp near backwater canal of Killai some 20 years back. It exactly resembles paddy in all characters and at that time it was considered as wild saline soil paddy.

4. *Cynodon dactylon (L) Pers (Plate – 37)*

Unlike normal short spreading plants, this is a longer trailing branched plant in Portonova sea-shore. Flowers very rare.

5. *Dactyloctenium aegysticum P. Beauv. (Plate – 37)*

Like the previous plant, this has long prostrate branches rooting at the nodes with shorter aerial branches. Flowers present. It is a completely modified plant due to salinity.

6. *Sporobolus diander (Retz.) P. Beauv (Plate 38)*

Present on the banks of salt water ditches at Pichavaram. At the tip of each branch a panicle inflorescence present. It is tripinnate in arrangement; leaves spirally arranged, slightly broad and sharp tip. A cluster of roots produced below and aerial branches above.

7. *Zoisia matrella (L)* **Merr** *(Plate 38)*

It is also present on the border of salt water ditches near sea shore at Pichavaram. A colony of plants thickly interwoven forming a thick mat wherever it is present. There is a horizontal main stem, from which a series of lateral branches are produced above. At the tip of each branch one or two inflorescence shoots arise. At the tip of each there is pinnately arranged spikelets on either side of the main axis. Each flower has an ovary with two stigma and 3 stamens.

8. *Arundo donax Linn (Plates 39 and 39a) Var. lagopoides var. nov D. Subramanian*

Gamble (1957) has stated that it is large plant upto 10 feet high present near water. But, the author has not so far seen it in Tamil Nadu.

Recently, this plant is found to form a continuous thick isolated colonies in swamps by the sides of back water canal at Poondiankuppam near Cuddalore. The plants are robust, 4 to 5 feet long and the tip of the shoots decompound inflorescence about, 1 feet long are produced; spikelets laterally compressed, not jointed on the pedicels, 2 to 8 flowered leaves ensiform, amplexicaul, 8 to 24 inch long, 0.5 to 2 inch broad, panicles 9 to 24 inches long. Sometimes the cattle feed the tip of the plant.

Gamble has not mentioned its presence in salt water areas. Besides, the present plants are half short as the plants mentioned by him. Therefore, the author has given it a new varietal name "lagopoides".

Spinifex littoreus (Burm. F merr.) (Plate – 40) (*S. squarrosus* Linn)

It is present in all seashore areas of Tamil Nadu, in salt soil, particularly at Rameswaram, Pamban and Mandapam. The author has collected this plant near Chidambaram. It is dioecious. Male plants and female plants are separate. Plants palegrey or glancous, stems forming thickets; leaves channeled, upto 13 inches long or heads up to 6 inches diameter, of heads upto 13 inches diameter, spikelets hidden at the base of the bracts. It is called Ravanan meesai. It is a good sand binder.

Cyperaceae

1. *Cyperus arenarius retz. (Plate 31)*

Rhizome creeping and dichotomously branched; at node an aerial shoot 7 to 8 cm in height produced with 10 to 15 cylindrical leaves spirally arranged at axils of leaves, 2 to 3 flowering spikes produced, spikelets slightly compressed. It is present near backwater canals of Pichavaram.

2. *Fimbristylis cymosa R. Br. (Plate 31)*

It is present near backwater canals of Pichavaram. From the basal root-stalk, cluster of roots are produced below and a bunch of leaves above. The leaves are thick and needle like. From the center of the plant an inflorescence is present above.

3. *Fimbrystylis lagopoides sp. nov. D. Subramaniam (Plate 32)*

This species is differing from the previous species in the nature of basal stem leaves and simple and unbranched inflorescence. The spikelets are produced 3 to 4 at a node. Likewise there are 2 nodes at the apex of the flowering shoot. It is available at Vellar estuary at Portonovo saline soil.

4. *Cyperus portonovensis sp. nov. D. Subramanian (Plate 32)*

This is a small species available in the same locality like the previous species at Portonovo. At the axils of the leaves included within them, there is a spike inflorescence. It is not raised above the leaves. So, it is difficult to collect them. There are 7 to 8 spikelets in an inflorescence.

In most of the characters, it is distinct from all other species of Cyperus of South India.

5. *Fimbristylis spathacea Roth (Plate 33)*

This plant is present on the saline soil areas near back water canal of Poondiankuppam near Cuddalore. There are clusters of a large number of needle like leaves produced on all sides. In the axils of leaves 2 to 3 inflorescences arise above the level of the leaves.

6. *Finbrystylis polytrichoides A. Br. (Plate 34)*

This is present in saline banks of back water canals of Pichavaram. Cluster of leaves and inflorescence are produced from the root-stalk. The spike is very simple and single.

PLATES

Plate – 1   *Acrostichum aureum* – A plant

Plate – 2   *Acrostichum speciosum* - A plant

Plate - 3   Figs. 1 to 3 = *Derris scandens* Fig. 1 = A twig with fruits; 2 = An inflorescens 3 = A flower enlarged Figs. 4 & 5 = *Derris trifoliata* Fig. 4 = A twig with flowers Fig. 5 = Fruit cluster

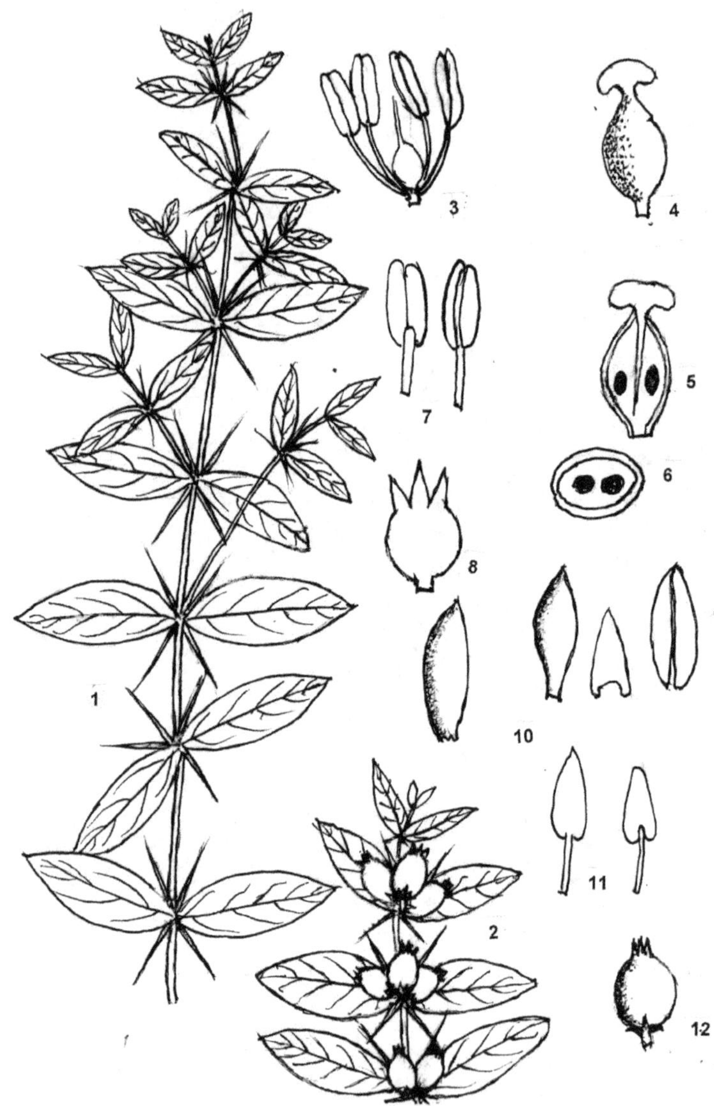

Plate – 4   *Azima tetracantha* Fig. 1 = A twig; 2 = A twig with fruits; 3 = Ovary & Stamens; 4 = Ovary; 5 = L.S. Ovary; 6 = T.S. Ovary 7 = Stamens; 8 = Calyx 9 = Corolla; 10 – Bract and Bracteoles; 11 = Staminode; 12 = Fruit

Plate – 5 *Salvadora Persica* Fig. 1 = A twig with inflorescenses; 2 = A flower = 3 = Calyx; 4 = Corolla; 5 = Petals with stamens; 6 = Stamens; 7 = T.S. and L.S. Ovary; 8 = A Fruit

Plate – 6   *Lumnitzera racemosa* Fig. 1 = A twig with flowers; 2 = Petal; 3 = Stamens; 4 = Flower; 5 = Stamens; 6 = Ovary L.S; 7 = Ovary T.S.; 8 = Floral diagram.

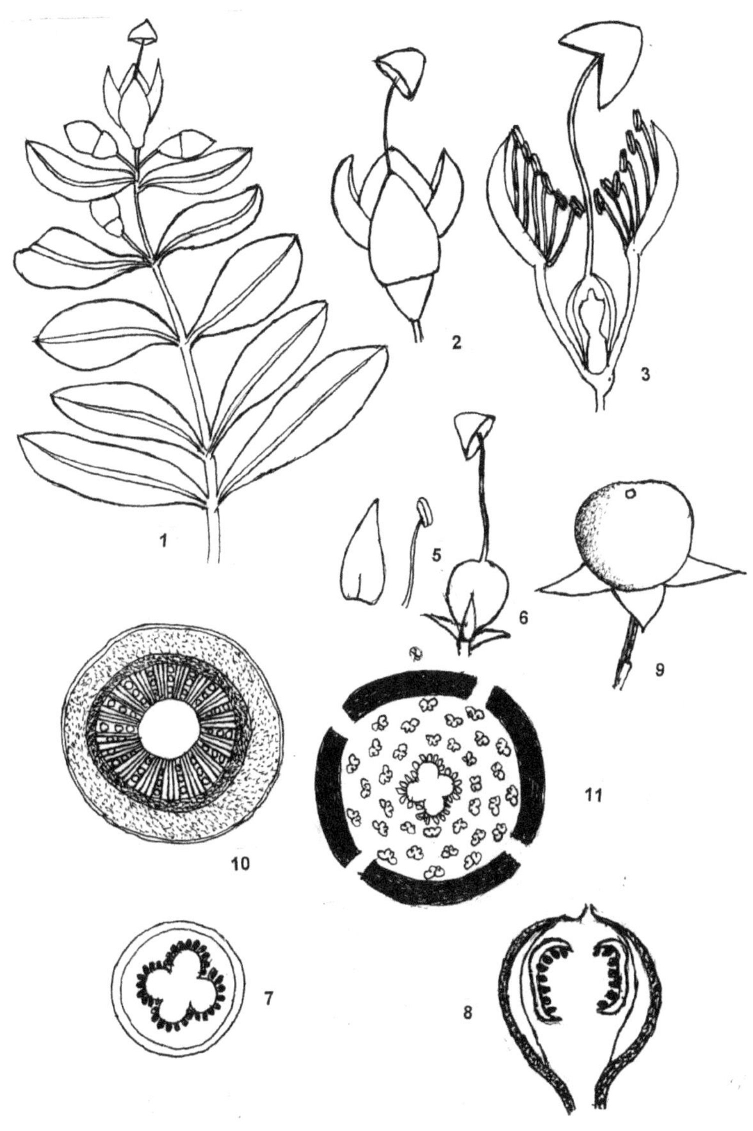

Plate – 7 *Sonneratia apetala* Fig. 1 = A twig with flowers; 2 = Flower; 3 = L.S. Flower; 4 = Sepal; 5 = Stamen; 6 = Ovary; 7 = T.S. Ovary; 8 = L.S. ovary; 9 = A fruit; 10 = T.S. Stem; 11 = Floral diagram.

Plate – 8 *Bruigiera cylindrica* Fig. 1 = A twig with flowers; 2 = Flower; 3 = Sepal; 4 = Petal; 5 = Stamen; 6 = T.S. and L.S. Ovary; 7 = Fruit; 8 = Floral diagram.

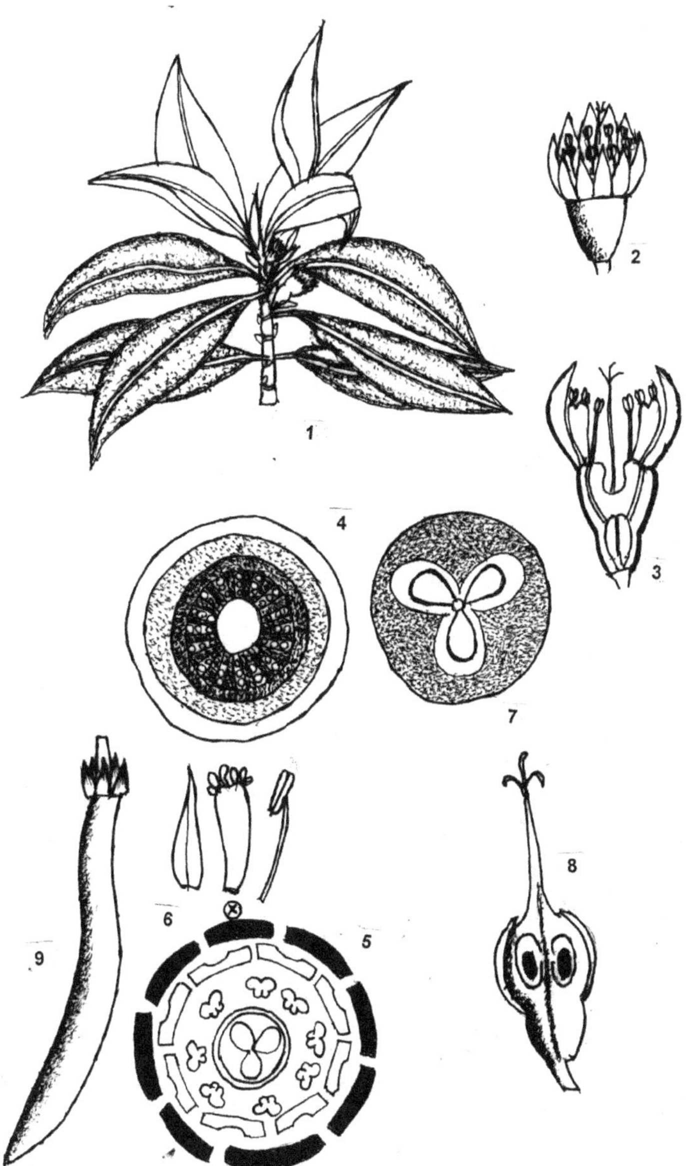

Plate – 9   *Bruigiera gymnorrhiza* Fig. 1 = A twig with flowers; 2 = A flower; 3 = L.S. Flower; 4 = T.S. Stem; 5 = Floral diagram; 6 = Sepal, Petal and Stamen; 7 = T.S. Ovary; 8 = L.S. ovary; 9 = A fruit.

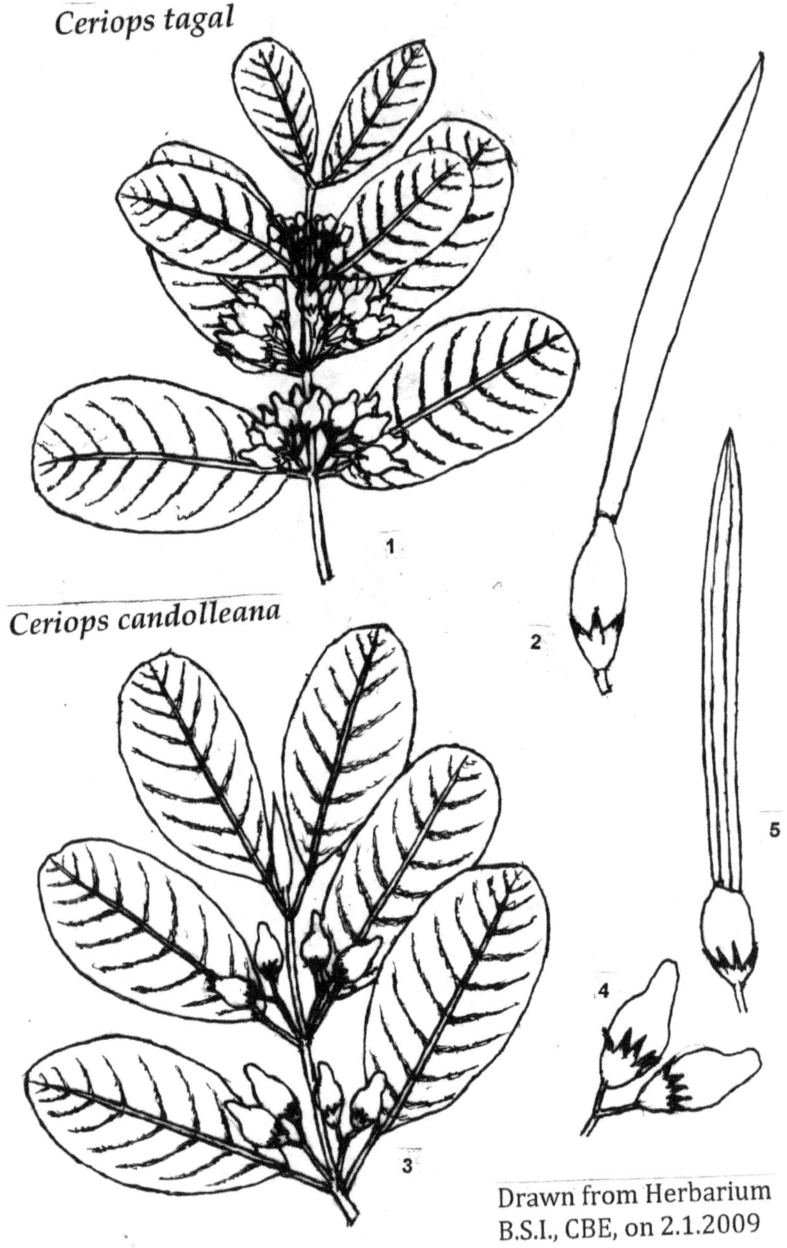

Plate – 10 Figs. 1 & 2 = Ceriops tagal; Fig. 1 = A twig; 2 = A fruit; Figs. 3 to 5 = Ceriops candolleana; Fig. 3 = A twig; 4 = Young fruits; 5 = A mature fruit.

Plate – 11 *Ceriops decandra* Fig. 1 = A twig; 2 = A flower; 3 = bract; 4 = Sepal; 5 = Petal; 6 = Stamen; 7 = Ovary; 8 = T.S. and L.S. Ovary; 9 = Young Fruit; 10 = Fruits; 11 = Floral diagram.

Plate – 12   *Rhizophora mucronata* Fig. 1 = A twig; 2 = Flower; 3 = Calyx; 4 = Petal; 5 = Sepal; 6 = Pistil; 7 = T.S. and L.S. Ovary; 8 = Stamens; 9 = A fruit

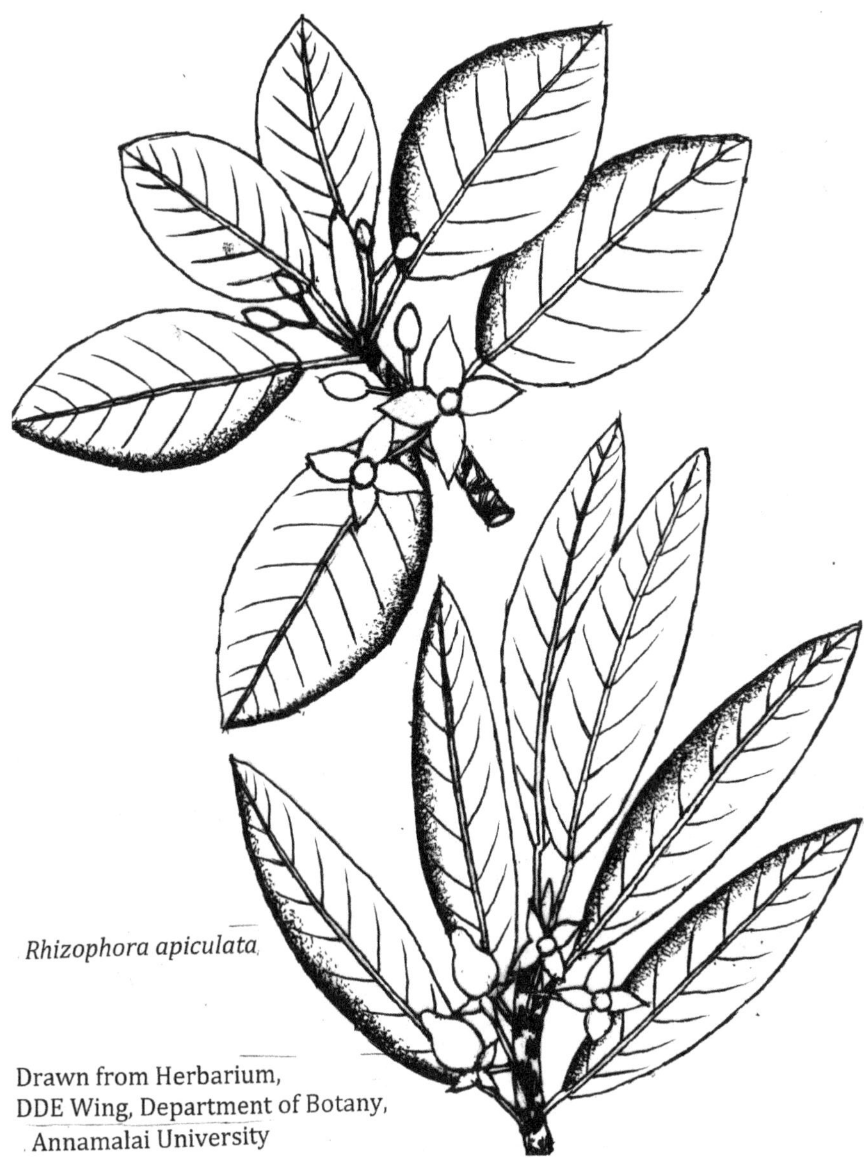

Plate – 13 *Rhizophora species*

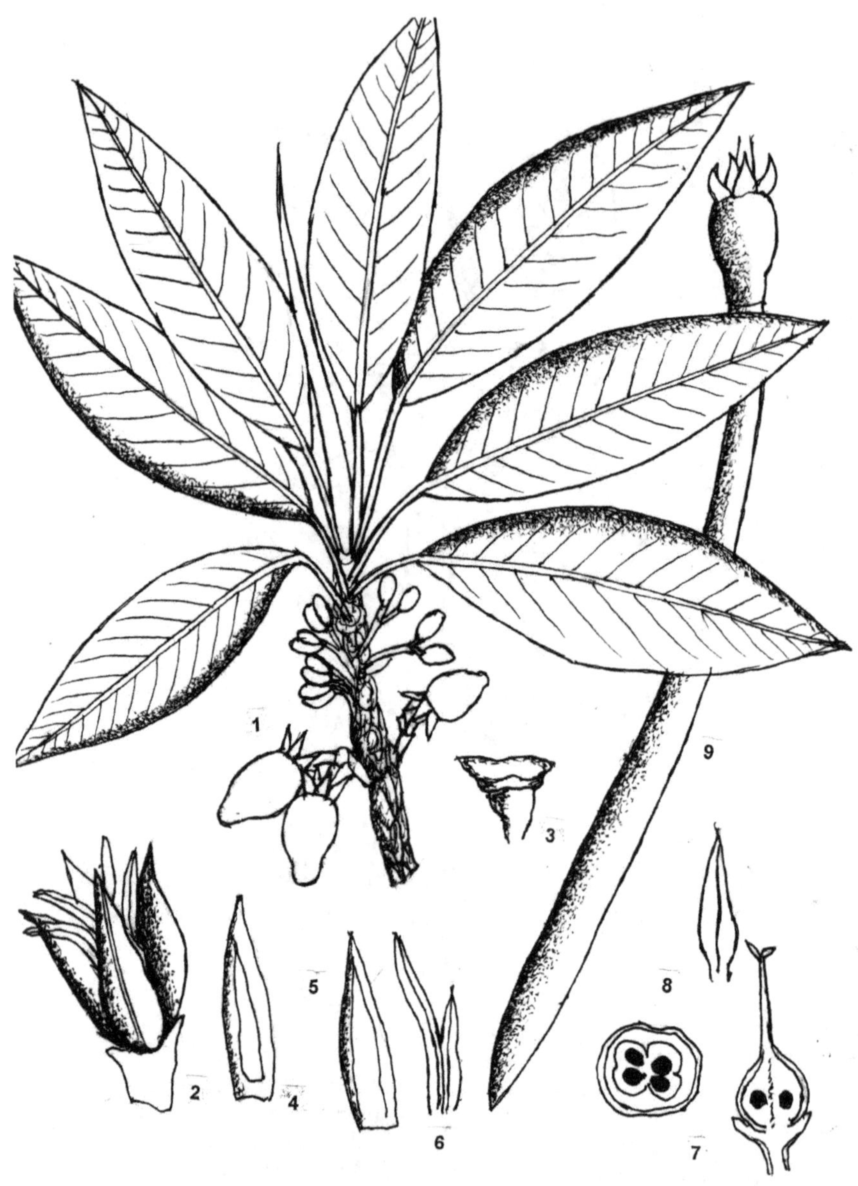

Plate – 14 *Rhizophora apiculata* Fig. 1 = A twig; 2 = A flower; 3 = Bracteoles; 4 = Sepal; 5 = Petal; 6 = Stamen; 7 = T.S. and L.S. Ovary; 8 = Stamen; 9 = Fruit.

Plate – 15 *Rhizophora – a peculiar population*

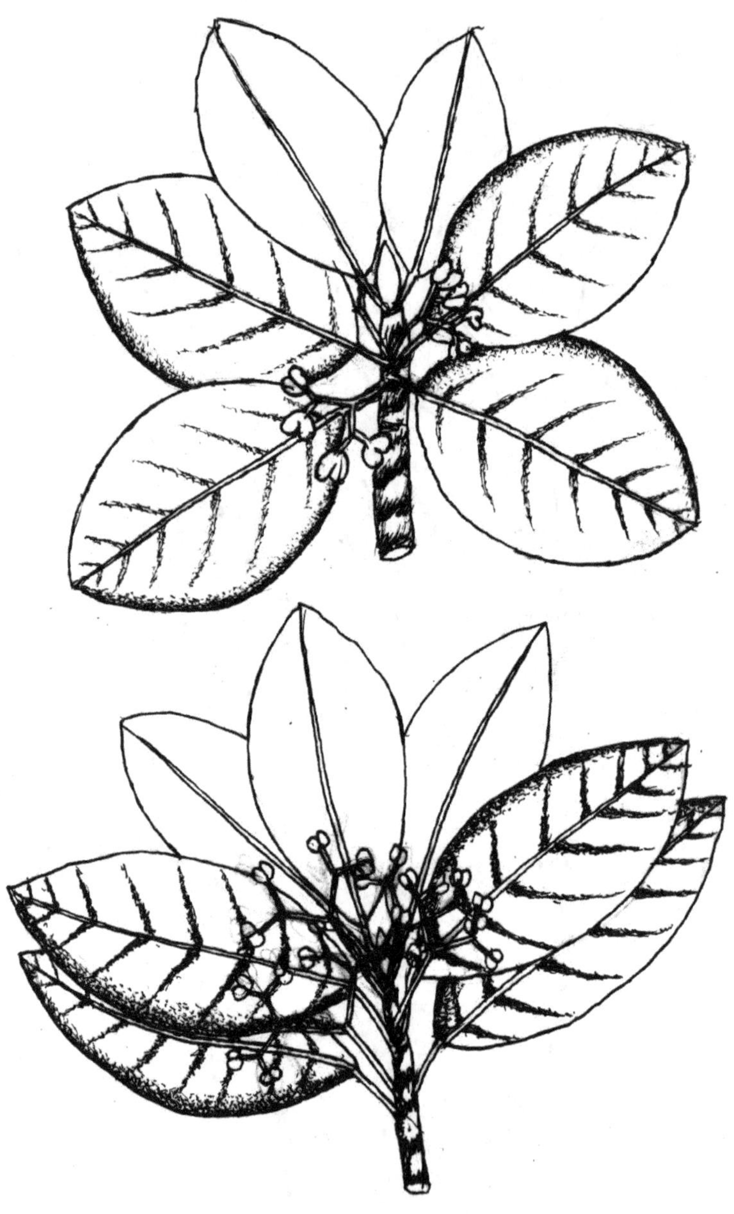

Plate – 16 *Rhizophora hybrid (R. apiculata x R. mucronata)*

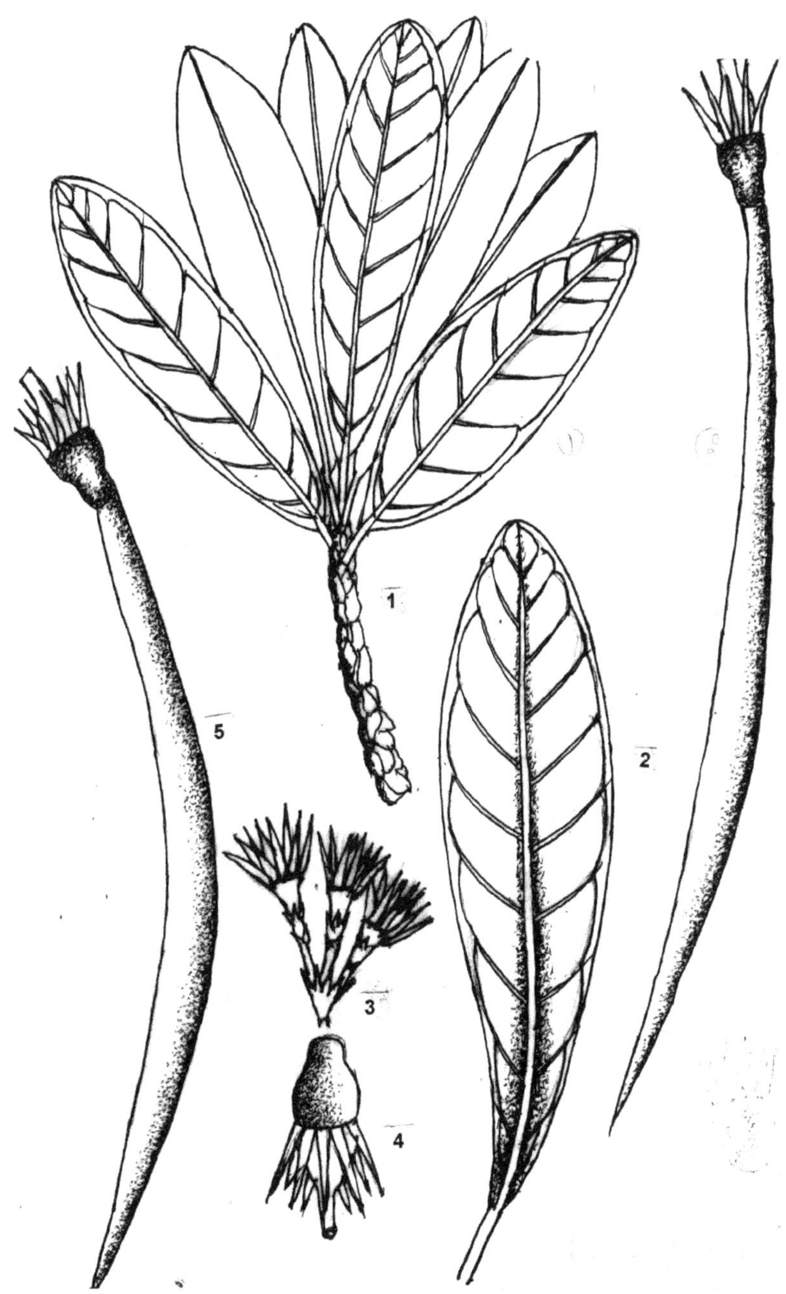

Plate – 17   *Kandelia candel* Fig. 1 = A twig; 2 = A leaf; 3 = Flowers; 4 = Young fruit; 5 = Fruit.

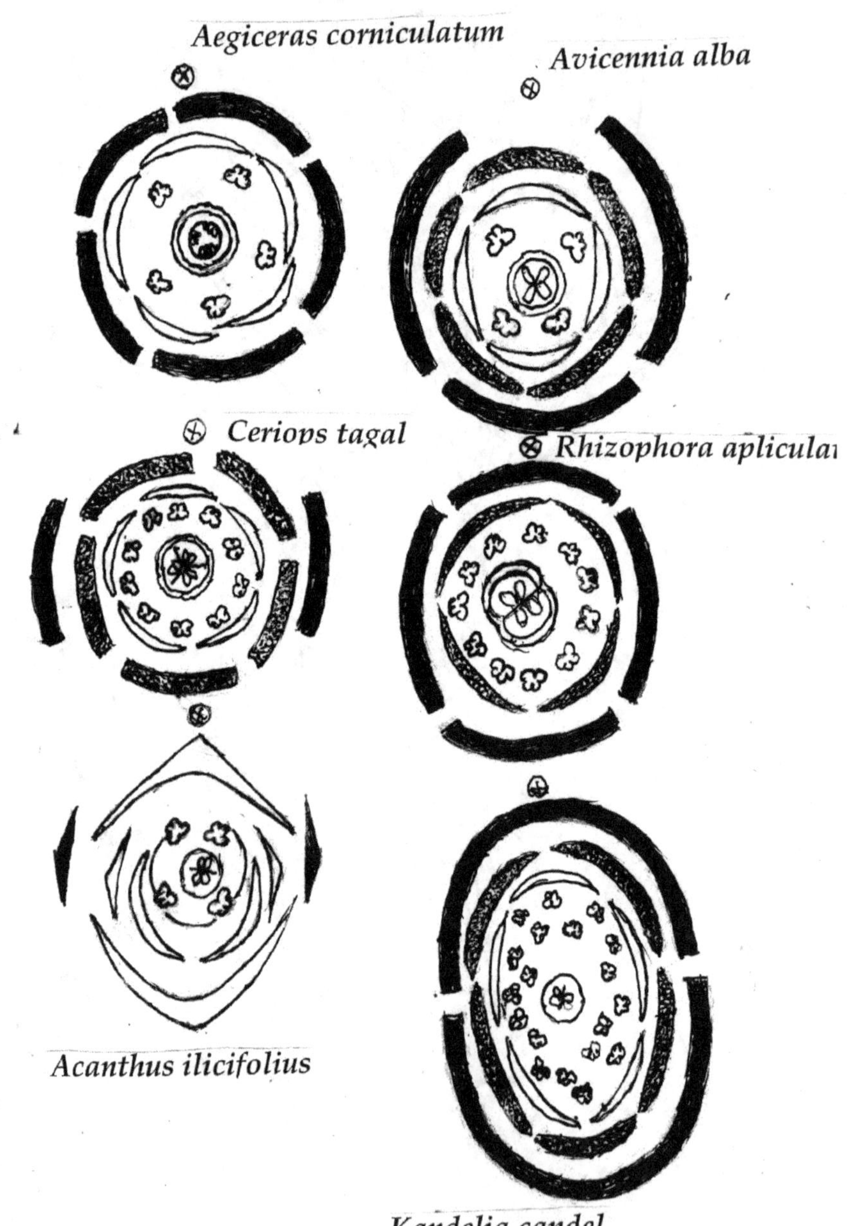

Plate – 18 *Floral diagrams*

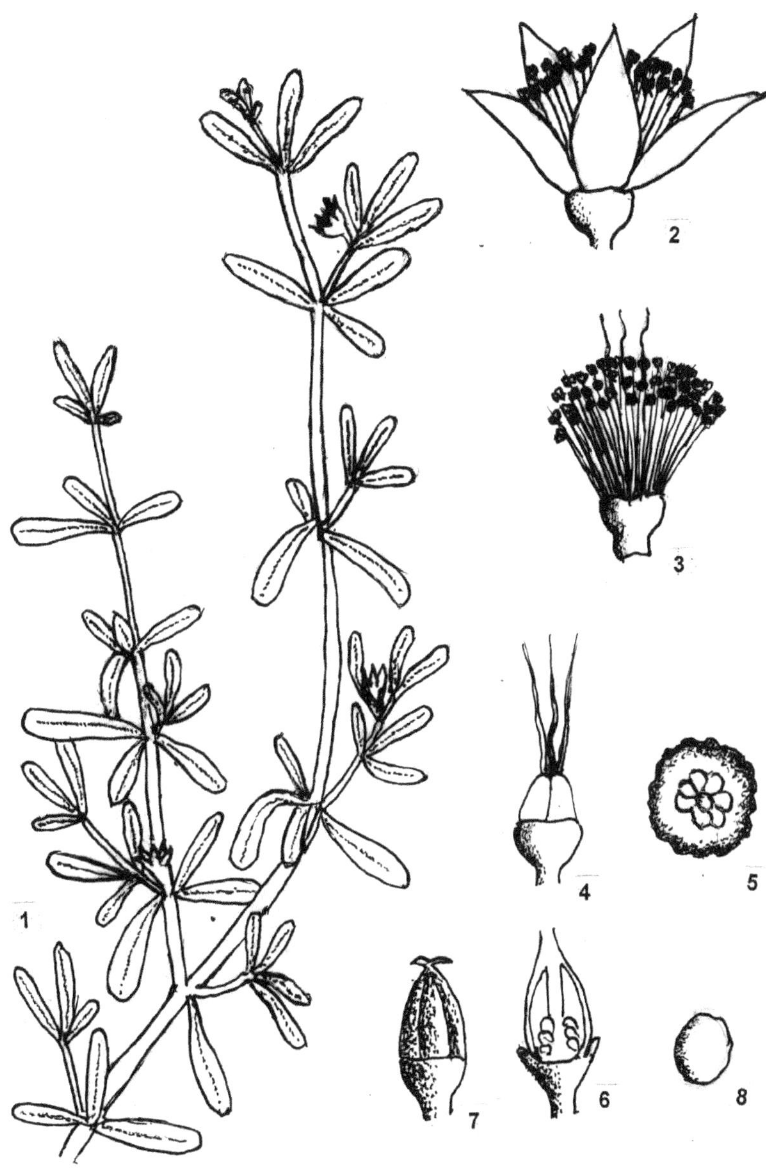

Plate – 19 *Sesuvium portulacastrum* Fig. 1 = A plant; 2 = A flower; 3 = A flower less petals; 4 = Ovary; 5 = T.S. Ovary; 6 = L.S. Ovary; 7 = A. Fruit; 8 = A seed.

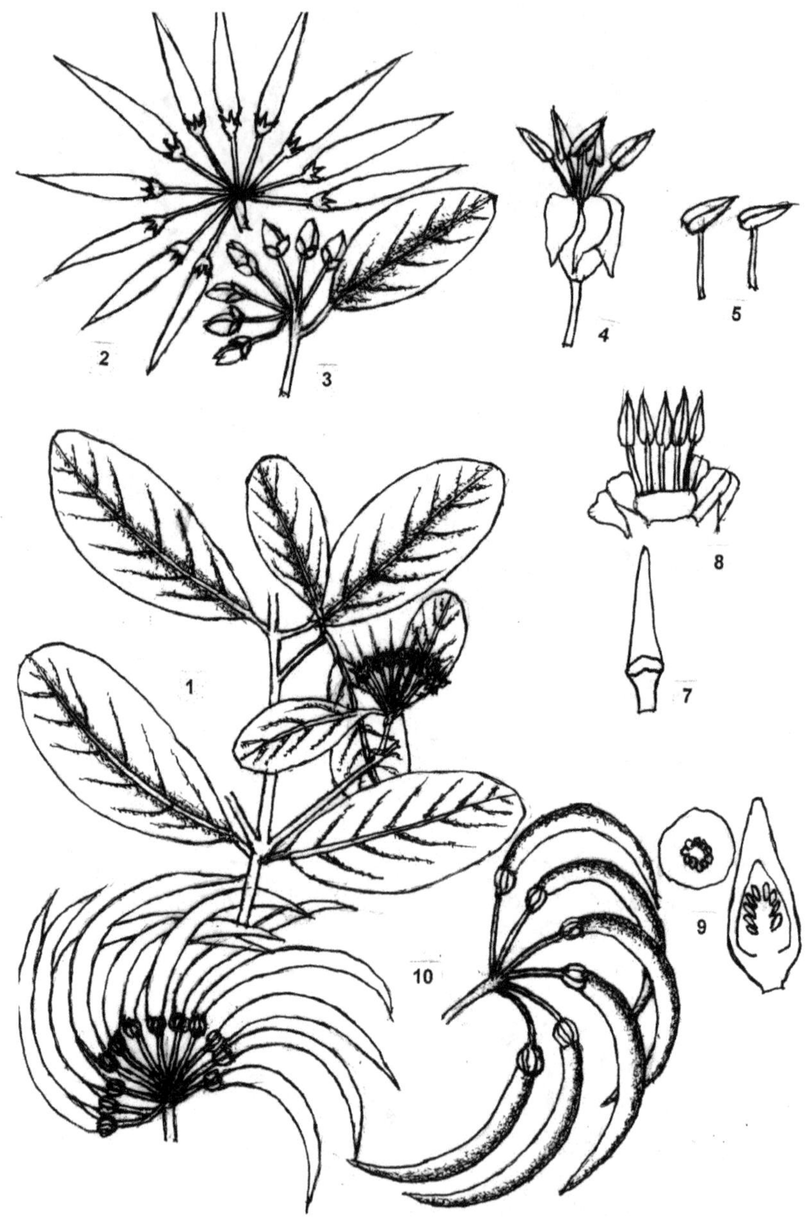

Plate – 20 *Aegiceros corniculatum* Fig. 1 = A twig; 2 = A young fruit cluster; 3 = Flower cluster; 4 = Flower; 5 = Stamen; 6 = Petals and Stamens; 7 = Ovary; 8 = T.S and L.S. ovary; 9 = Fruit Clusters.

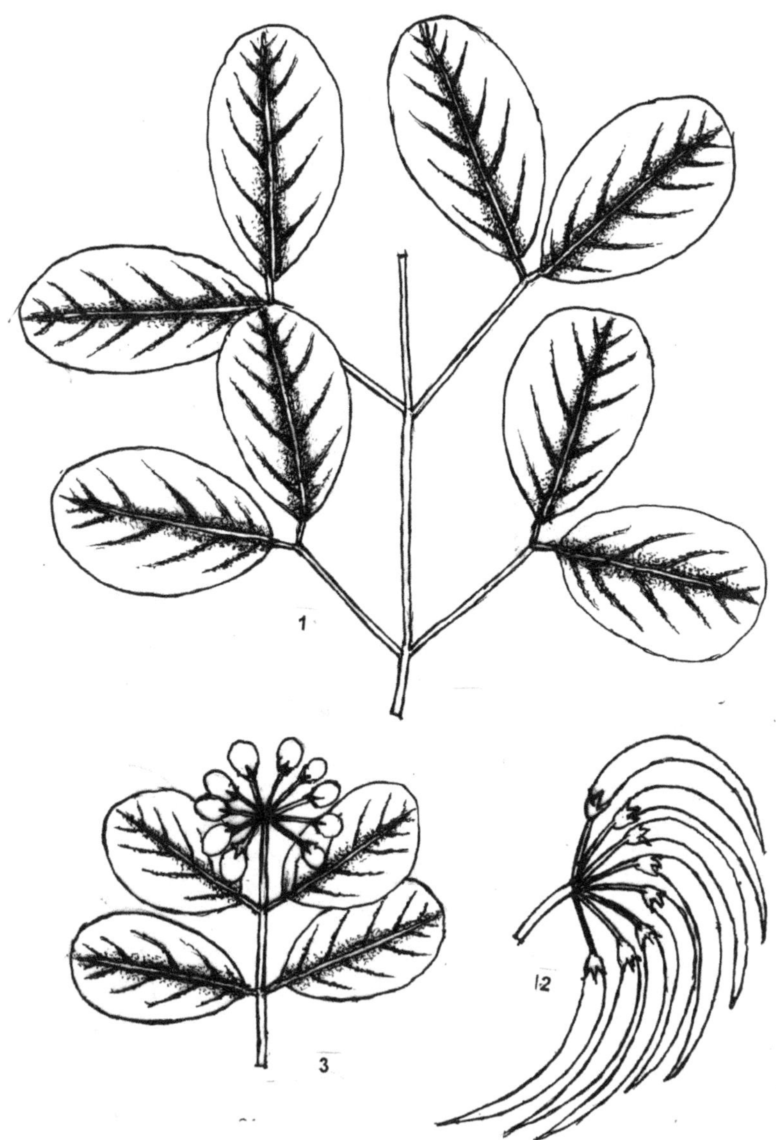

Plate – 21 *Aegiceros majus* Fig. 1 = A plant; from kille 2 = Fruit cluster; 3 = A plant from Cuddalore.

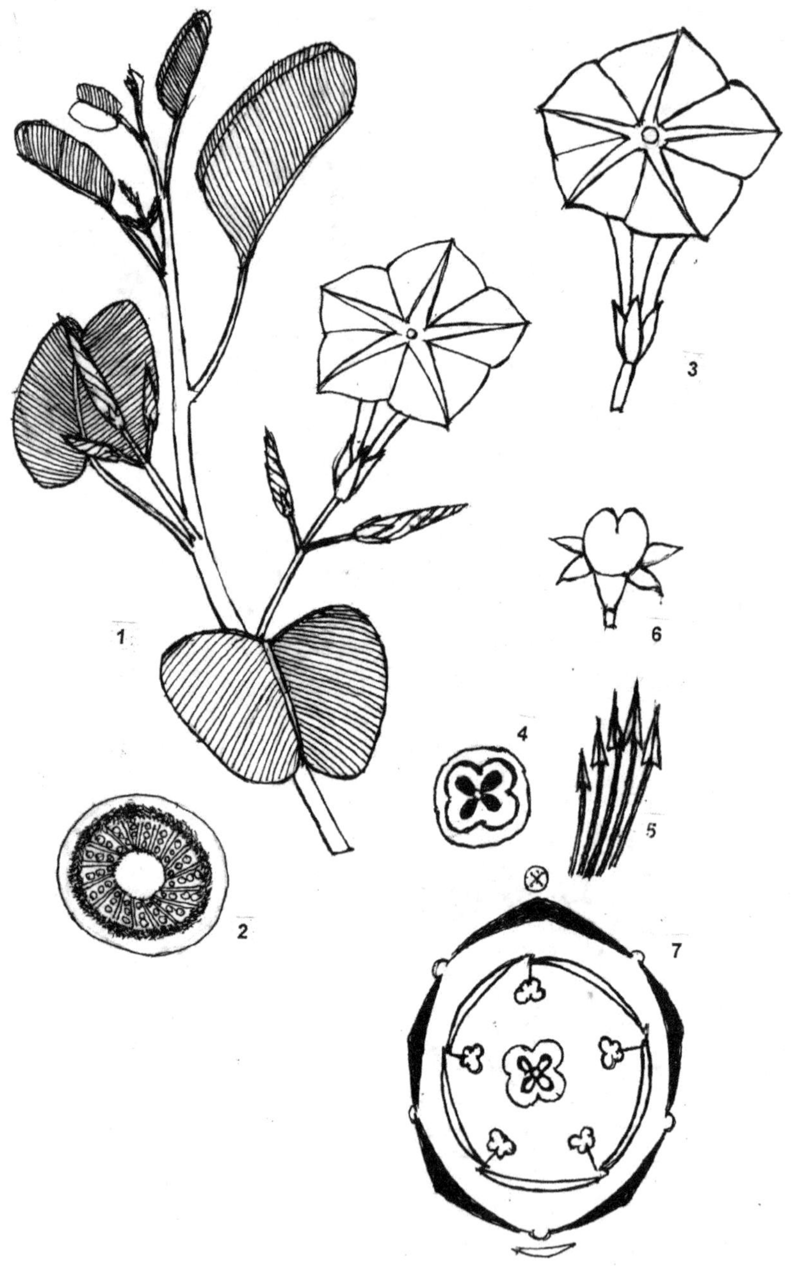

Plate – 22   *Ipomea biloba* Fig. 1 = A twig; 2 = T.S. stem, 3 = A flower; 4 = T.S. Ovary, 5 = Stamens; 6 = Fruit; 7 = Floral diagram.

Plate – 23   *Acanthus ilicifolius* Fig. 1 = A twig; 2. flower, side view; 3 = Flower, face view, 4 = Fruit cluster; 5 = Ovary; 6 = Sepal, petals; 7 = Stamens; 8 = A fruit.

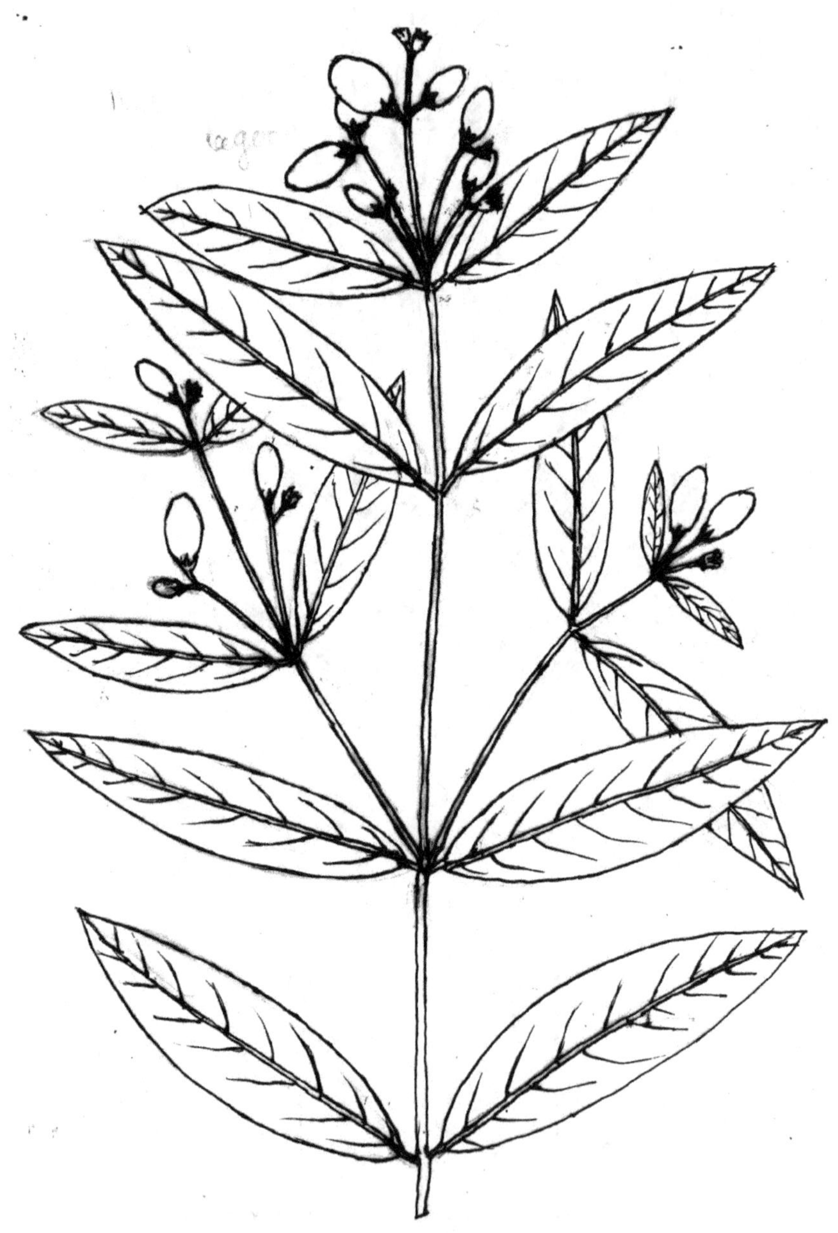

Plate – 24   Avicennia alba

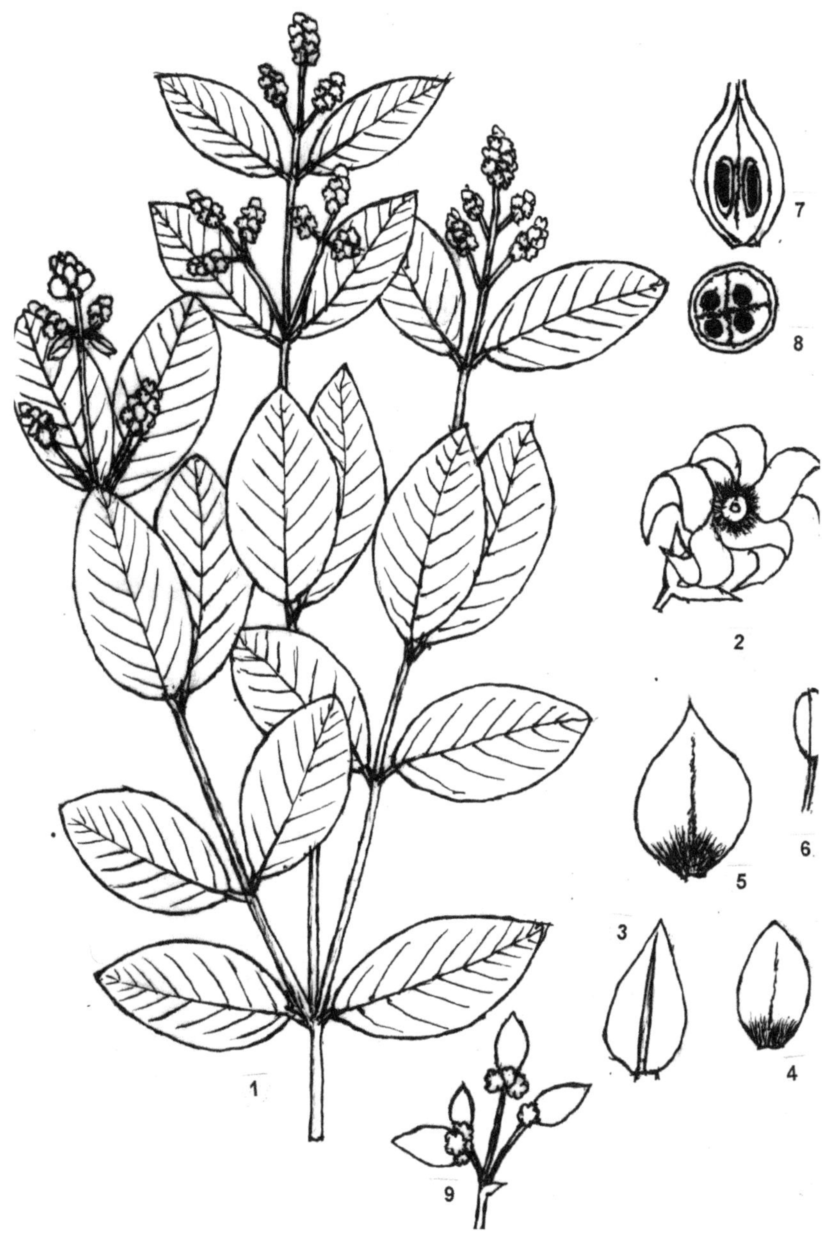

Plate – 25 *Avicennia marina* Fig. 1 = A twig; 2 = A flower; 3 = Sepal; 4 & 5 = Petals; 6 = Stamen; 7 = T.S. Ovary; 8 = L.S. Ovary; 9 = Fruit cluster.

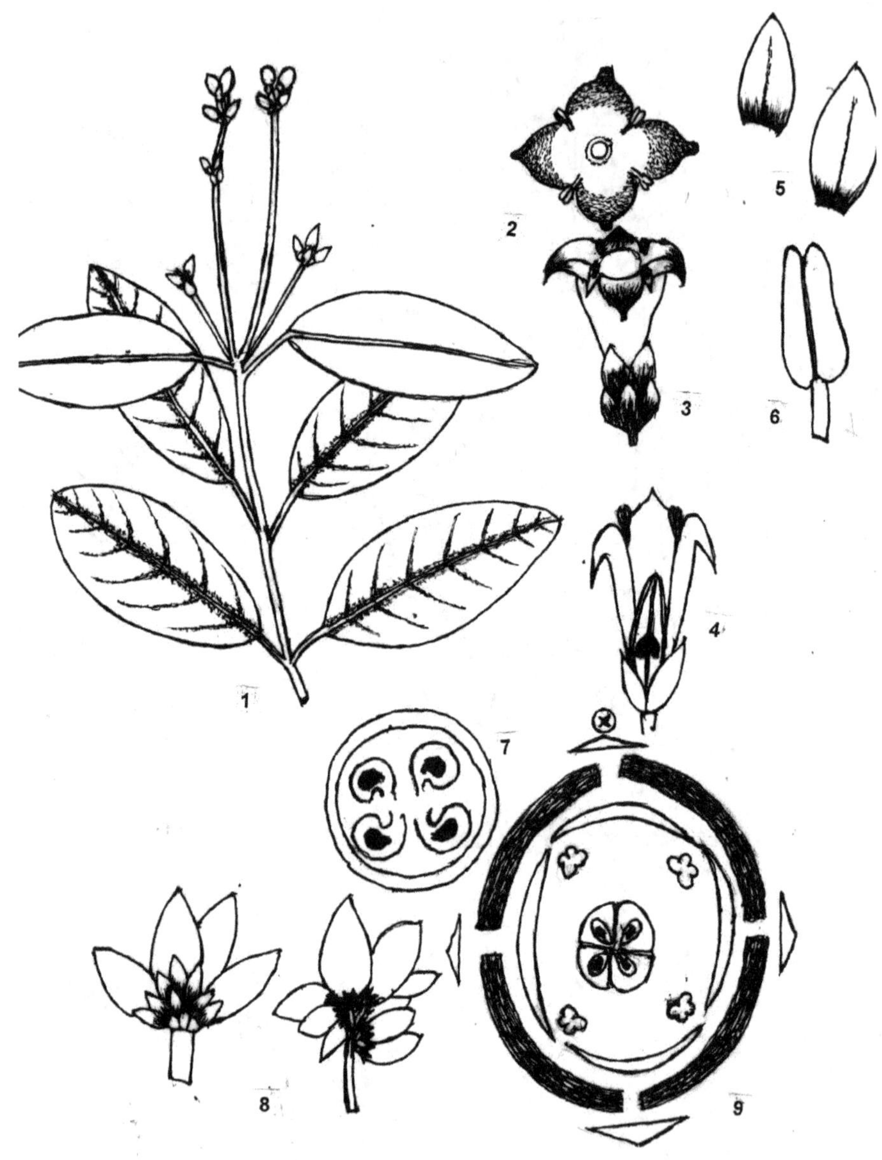

Plate – 26   *Avicennia officinalis* Fig. 1 = A twig; 2 = Flower face view; 3 = Flower, side view; 4 = L.S. Flower; 5 = Sepals & Petal; 6 = Stamen; 7 = T.S. Ovary; 8 = Fruit Clusters; 9 = Floral diagram.

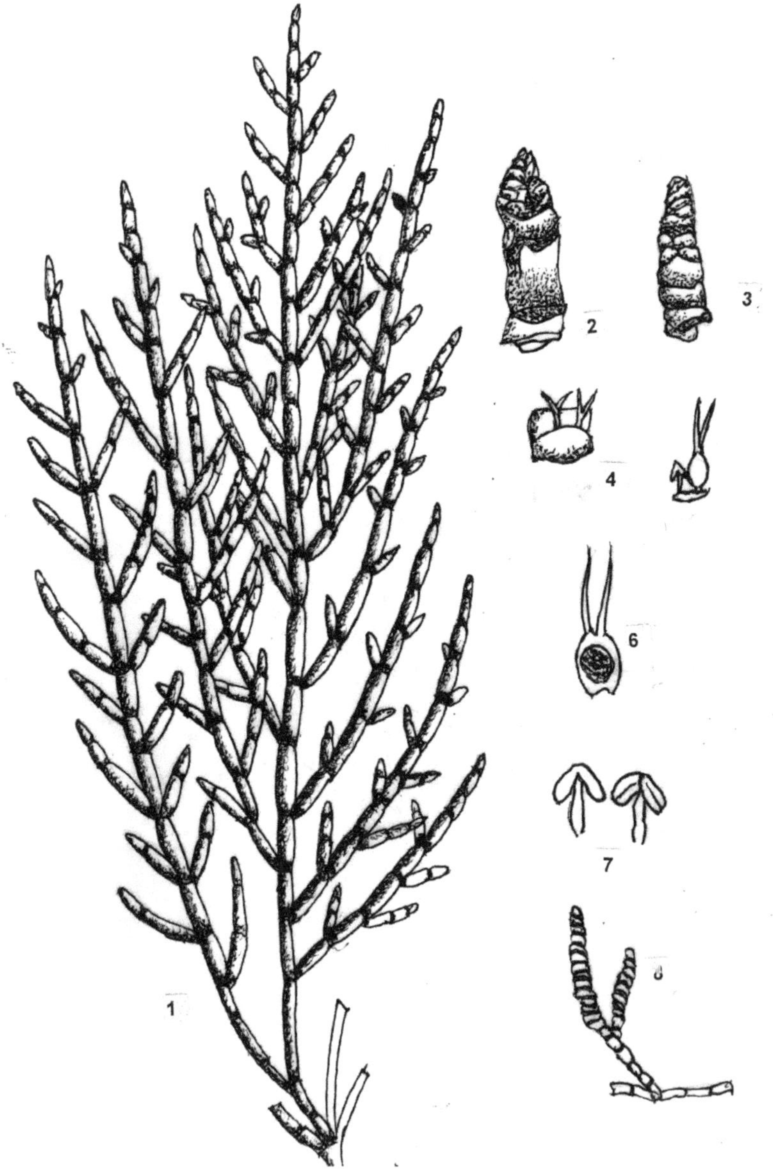

Plate – 27   *Arthrocnemum indicum* Fig. 1 = Vegetative shoot; 2 & 3 = Flowering shoots; 4 = Flowers; 5 = Ovary and Stamen; 6 = T.S. Ovary; 7 = Stamens; 8 = Condensed shoot.

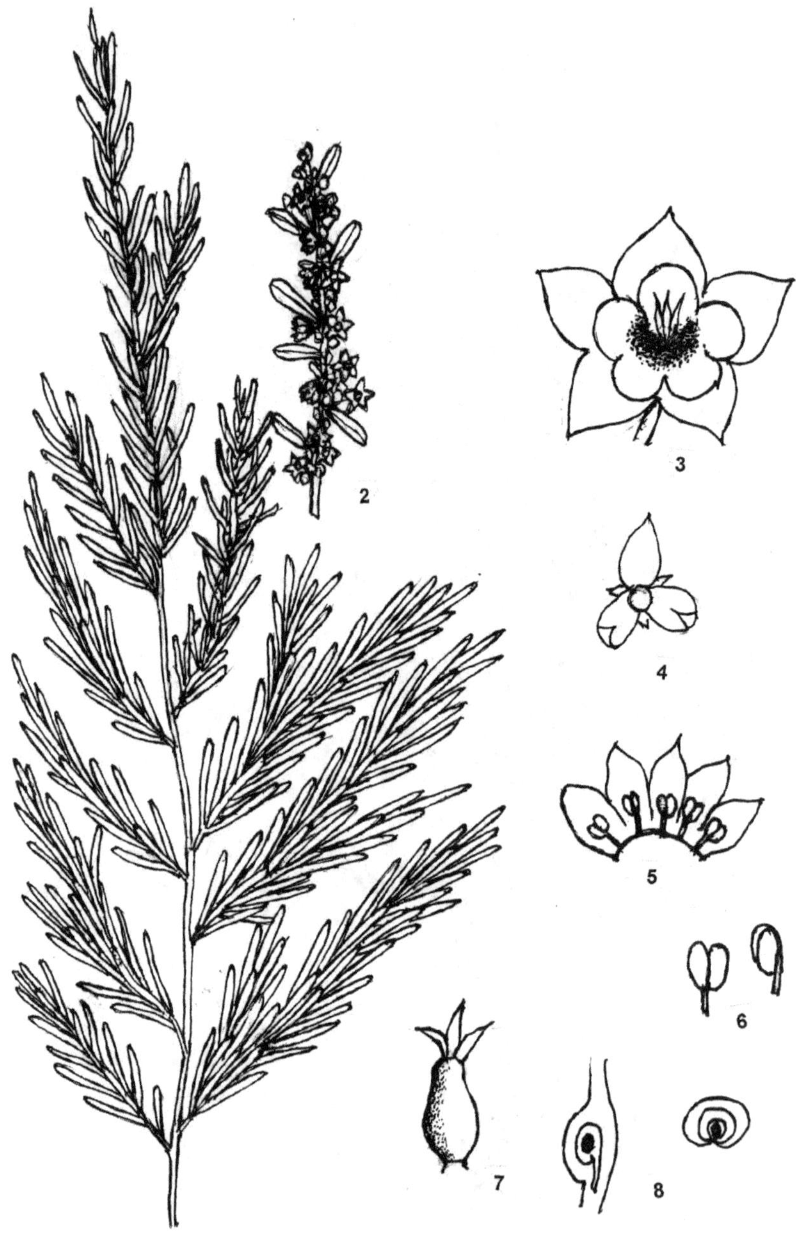

Plate — 28 *Suaeda monoica* Fig. 1 = Vegetative shoot; 2 = Flowering shoot; 3 = A flower; 4 = Bracts & Bracteoles; 5 = Petals & stamens; 6 = Stamens; 7 = Ovary; 8 = T.S. and L.S. Ovary.

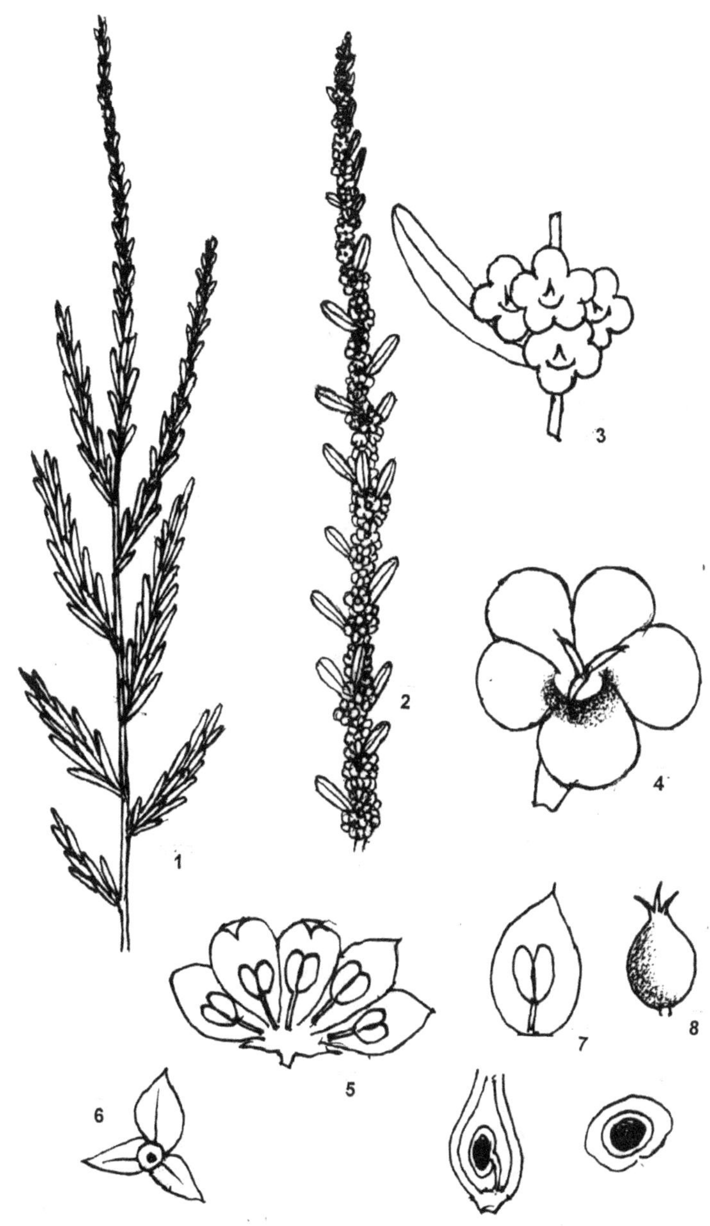

Plate – 29 *Suaeda nudiflora* Fig. 1 = Vegetative shoot; 2 = Flowering shoot; 3 = Flowers; 4 = Flower; 5 = Petals & Stamens; 6 = Bracts & Bracleoles; 7 = A petal and a Stamen; 8 = Ovary; 9 = T.S. and L.S. Ovary.

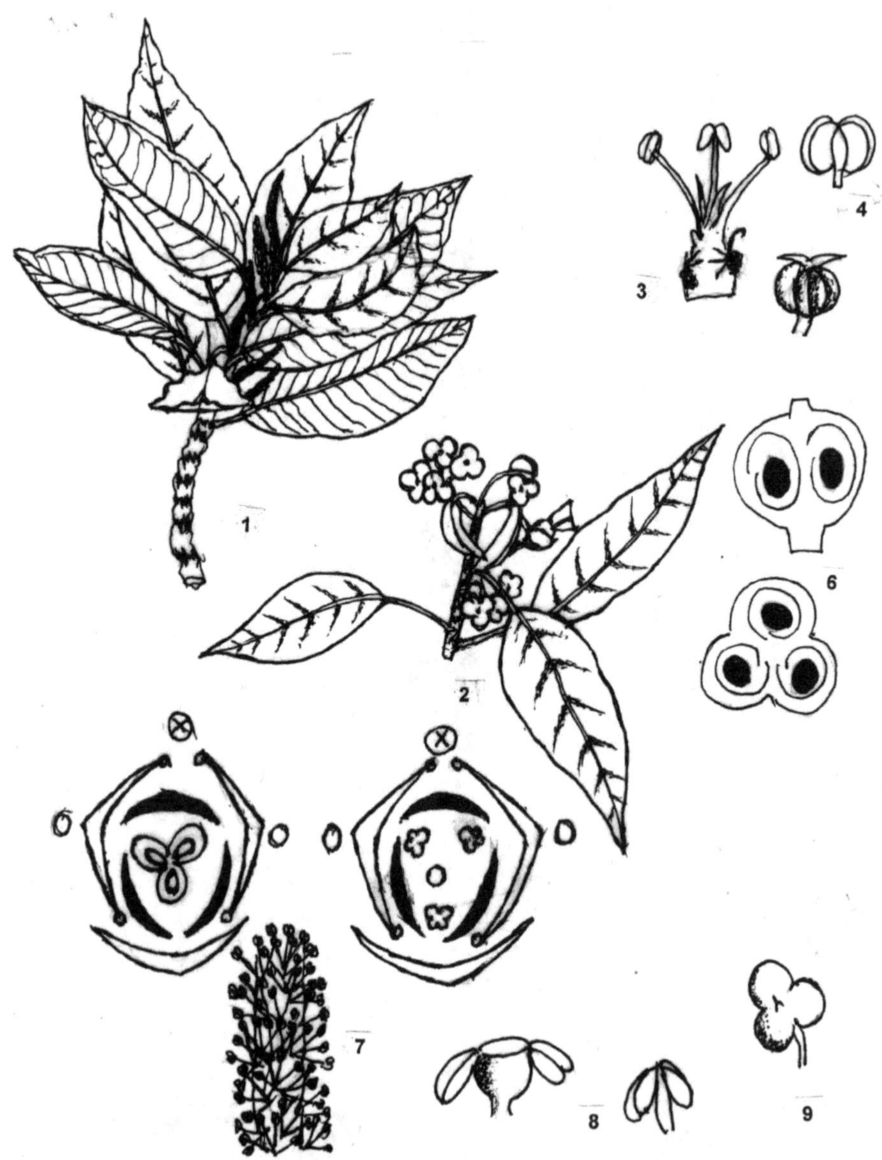

Plate – 30 *Excoecaria agallocha* Fig. 1 = A twig; 2 = A fruiting twig; 3 = A female flower; 4 = Stamens; 5 = Carpel; 6 = T.S. and L.S. Ovary; 7 = Male inflorecens; 8 = Stamens; 9 = Fruit; 10 = Floral diagrsm, male and female.

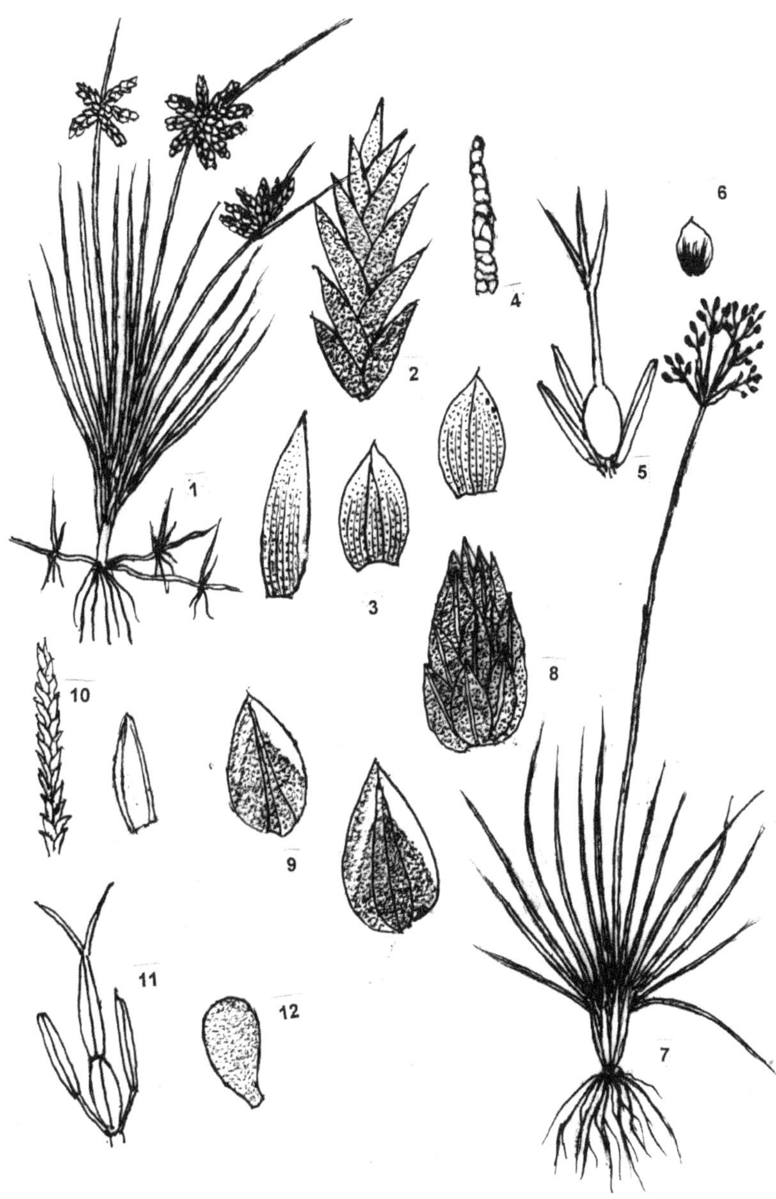

Plate – 31 Figs. 1 to 6 = *Cyperus arenarius*; Fig. 1 = A plant; 2 = Spike; 3 = Bract & sepals; 4 = Rachilla; 5 = A flower; 6 = Fruit; Figs. 7 to 12 = *Fimbristylis cymosa*; Fig. 7 = A plant; 8 = Spike; 9 = Glumes; 10 = Rachilla; 11 = A Flower; 12 = Fruit.

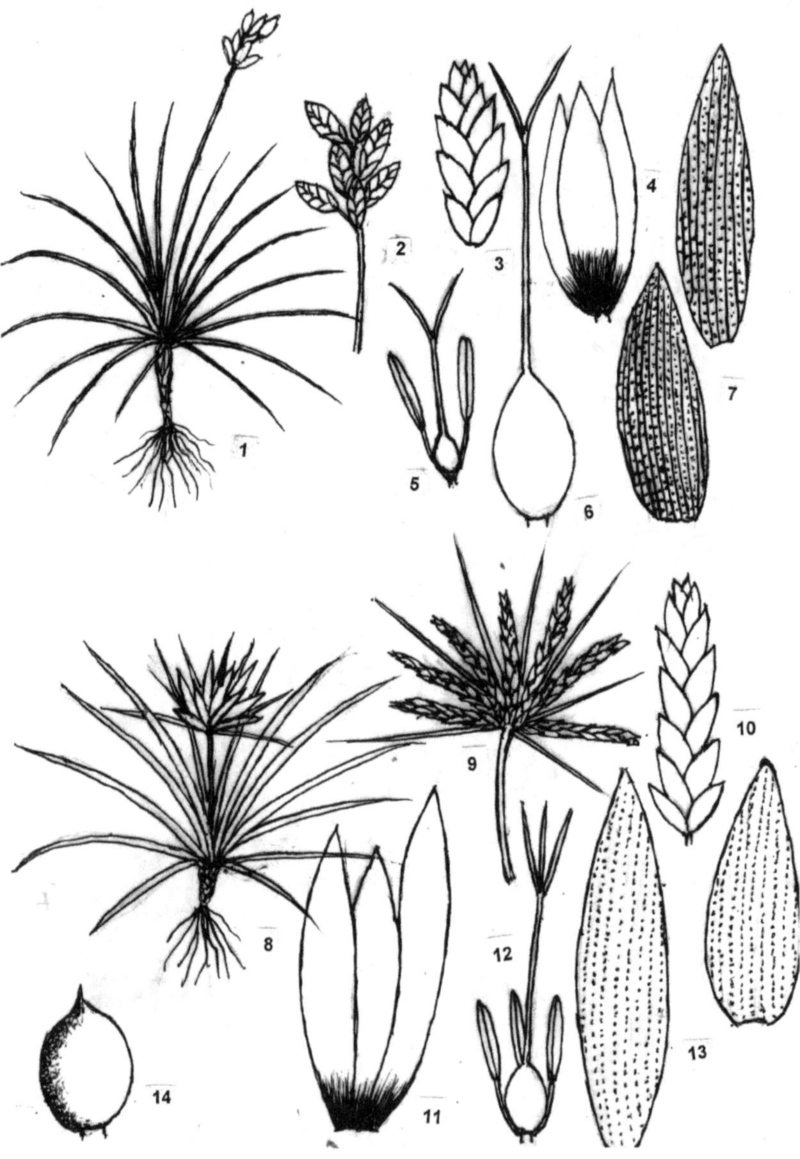

Plate – 32 Figs. 1 to 7 = *Fimbristylis lagopoides* Fig. 1 = A plant; 2 = Spikes; 3 = Spike enlarged; 4 = Flowers; 5 = A flower; 6 = Ovary; 7 = Glumes;' Figs. 8 to 13 = *Cyperus portonovensis*; Fig. 8 = A plant; 9 = Spikes; 10 = Spike enlarged; 11 = Spikelet; 12 = A flower; 13 = Glumes; 14 = Fruit.

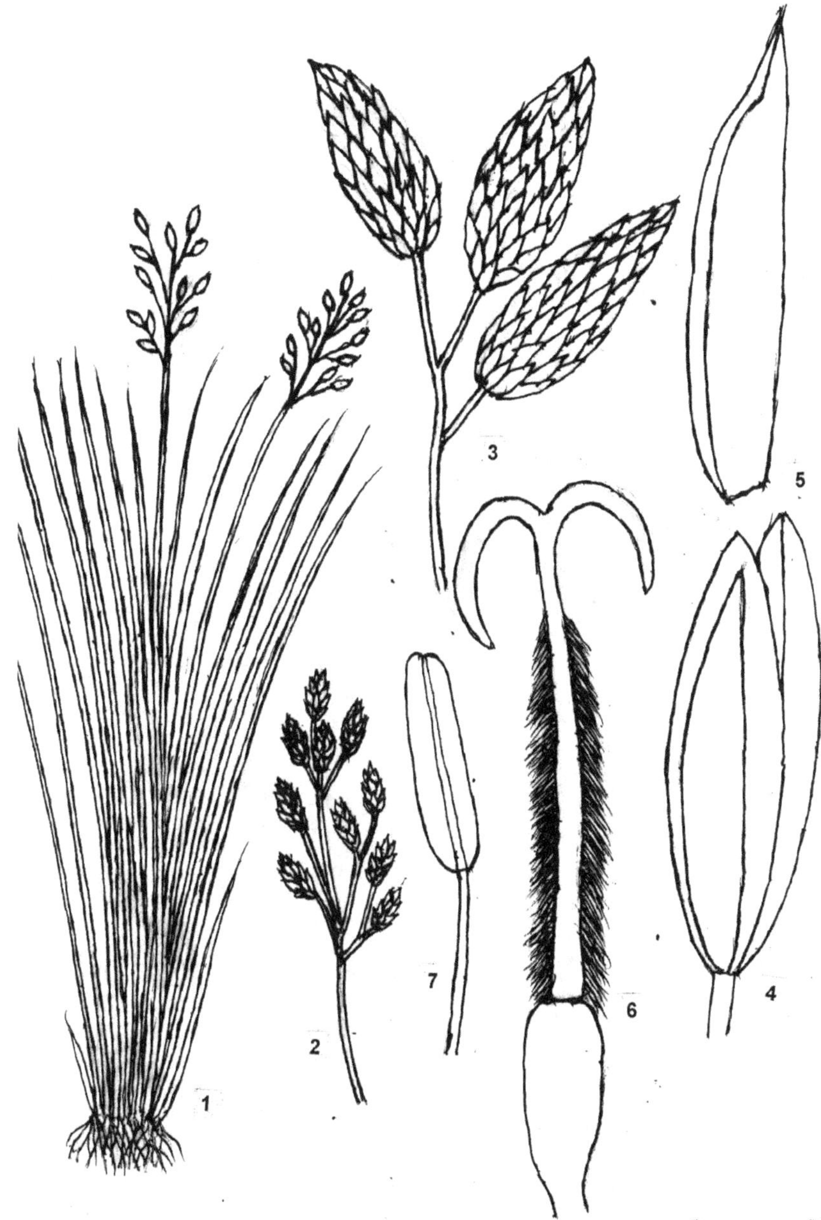

Plate – 33   *Fimbristylis spathacea* Fig. 1 = Plant; 2 = Spikes; 3 – Few spikes enlarged; 4 = Spikelet; 5 = Glume; 6 = Ovary; 7 = Stamen.

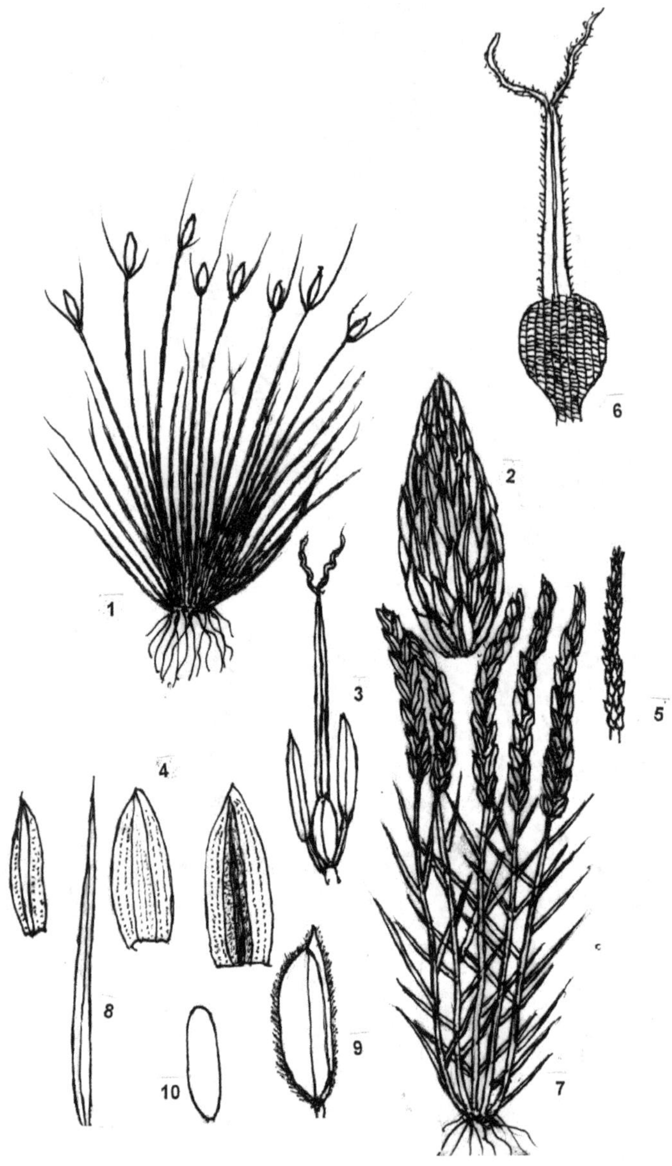

Plate – 34   Figs. 1 to 6 = *Fimbristylis polytrichoides* Fig. 1 = Plant; 2 = Spike enlarged; 3 = Flower, 4 = Bract sepal & petal; 5 = Rachilla; 6 = Ovary; Figs. 7 to 10 = *Portiresia coarctata*; Fig. 7 = Plant; 8 = Leaf; 9 = Fruit; 10 = Seed.

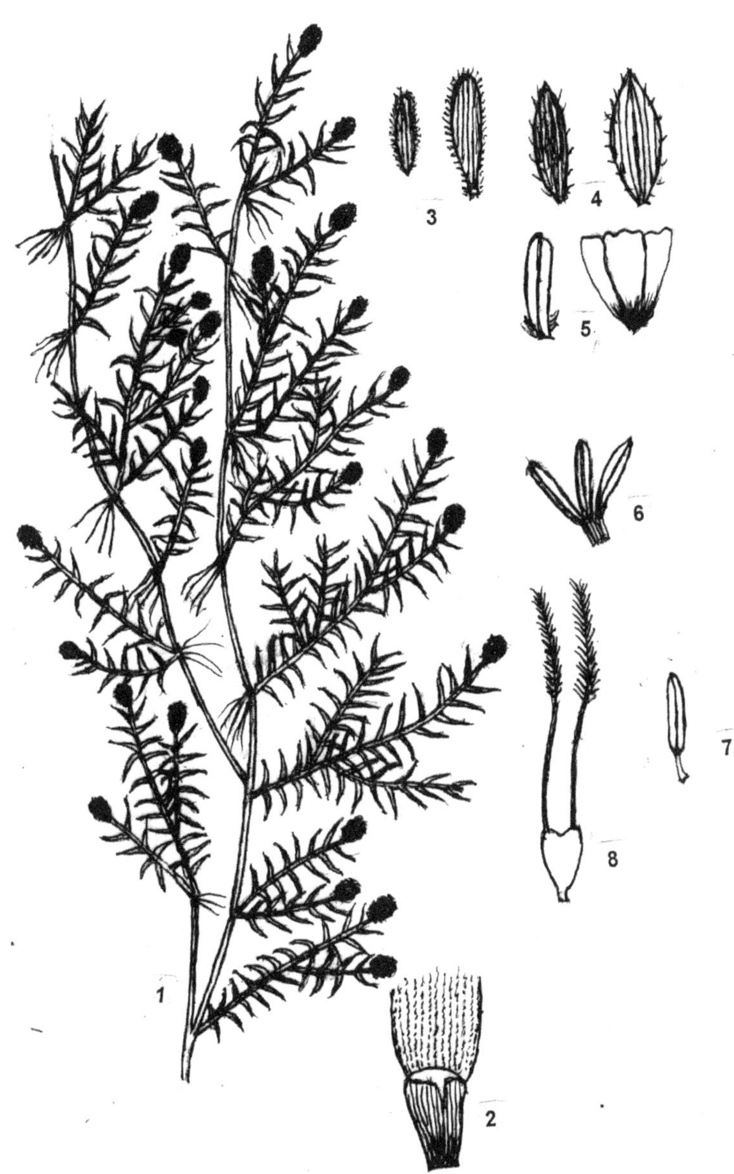

Plate - 35   *Aeluropus lagopoides* Fig. 1 = Plant; 2 = Glume; 3 = Lower and Upper glumes; 4 = Flowering glumes; 5 = Limb; 6 = Stamens; 7 = Stamen; 8 = Ovary

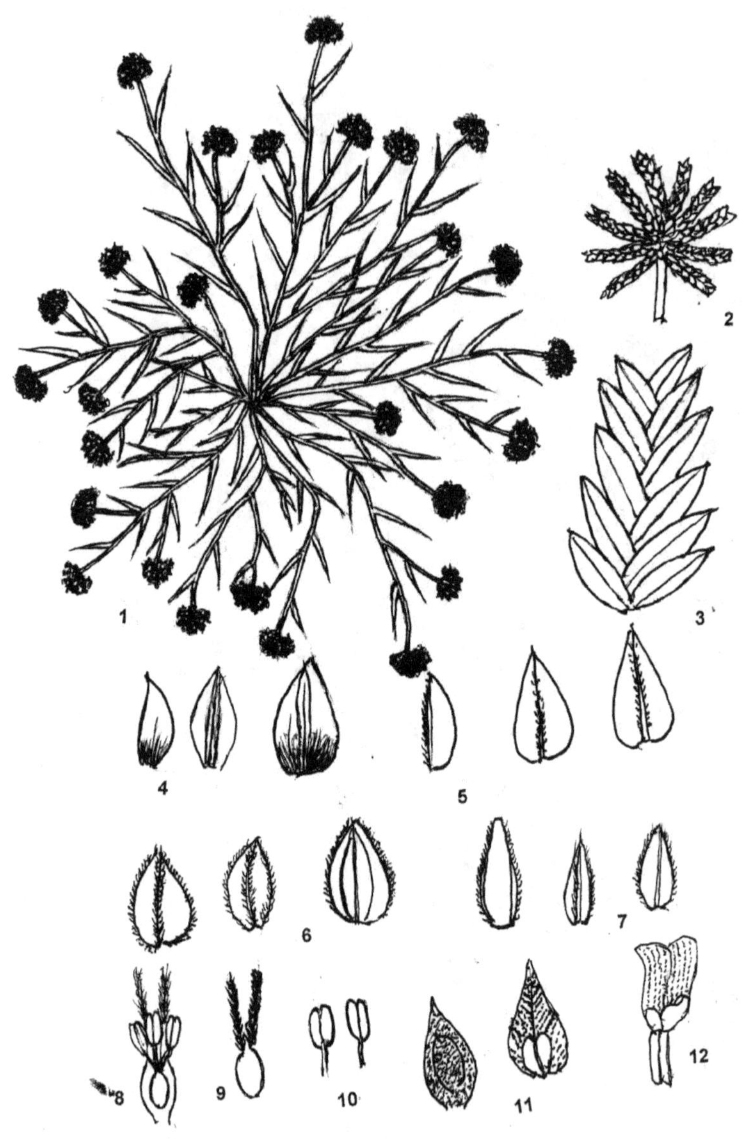

Plate – 36 *Coelachryum lagopoides* Fig. 1 = Plant; 2 = Spike cluster; 3 = Spike; 4 = Upper glumes; 5 = Flowering glumes; 6 = Lemma; 7 = palaea; 8 = Flower; 9 = Ovary; 10 – Stamens; 11 = Caryopsis; 12 = Limb.

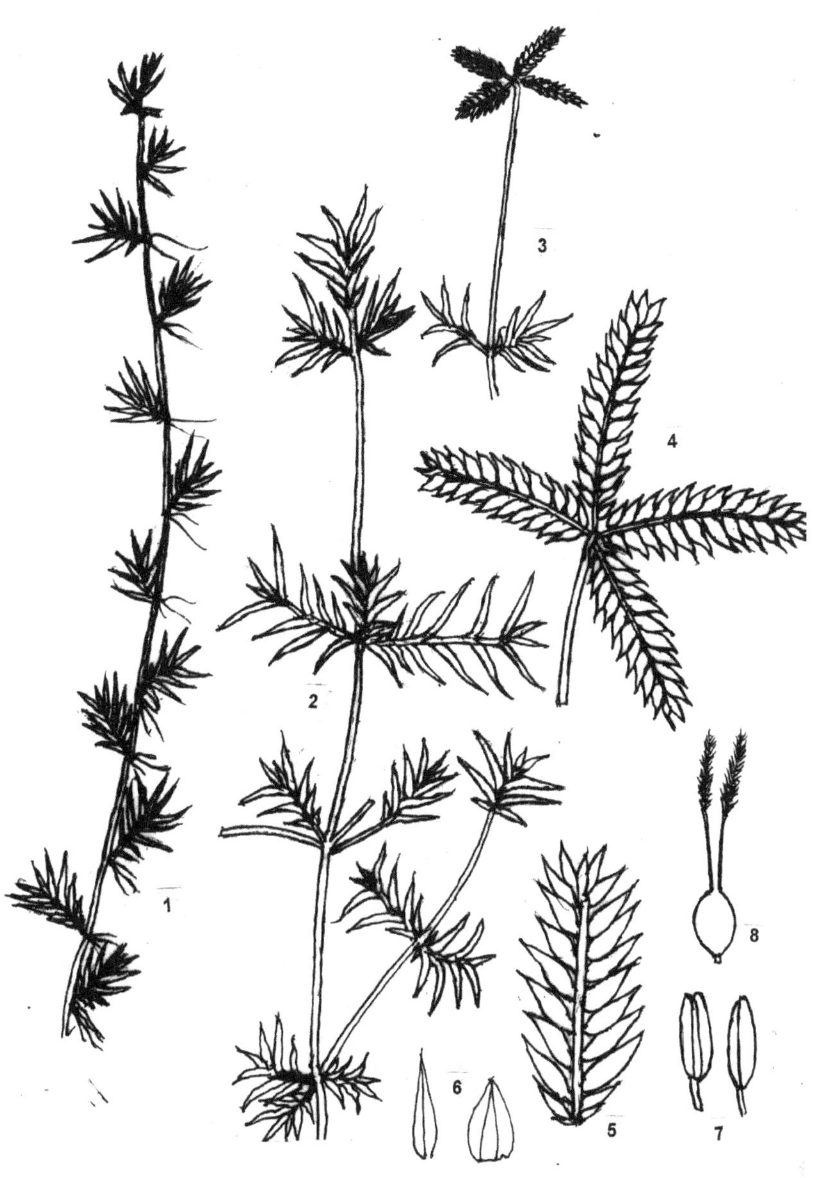

Plate – 37 Fig. 1 = *Cynodon dactylon* Long prostrate twining and crawling branch. Fig. 2 to 7 = *Dactyloctenium aegypticum* Fig. 2 = Long prostrate crawling branch; 3 = Plant tip with spikes; 4 = Spikes enlarged; 5 = A spike; 6 = Glumes; 7 = Stamens; 8 = Ovary.

Plate – 38   Figs. 1 to 5 = *Sporobolus diander* Fig. 1 = A plant; 2 = Ligule; 3 = Upper and Lower Glumes; 4 = Lemma; 5 = Flower and Palea Figs. 6 to 8 = *Zoisia matrella* Fig. 6 = A plant; 7 = A flower; 8 = A fruit.

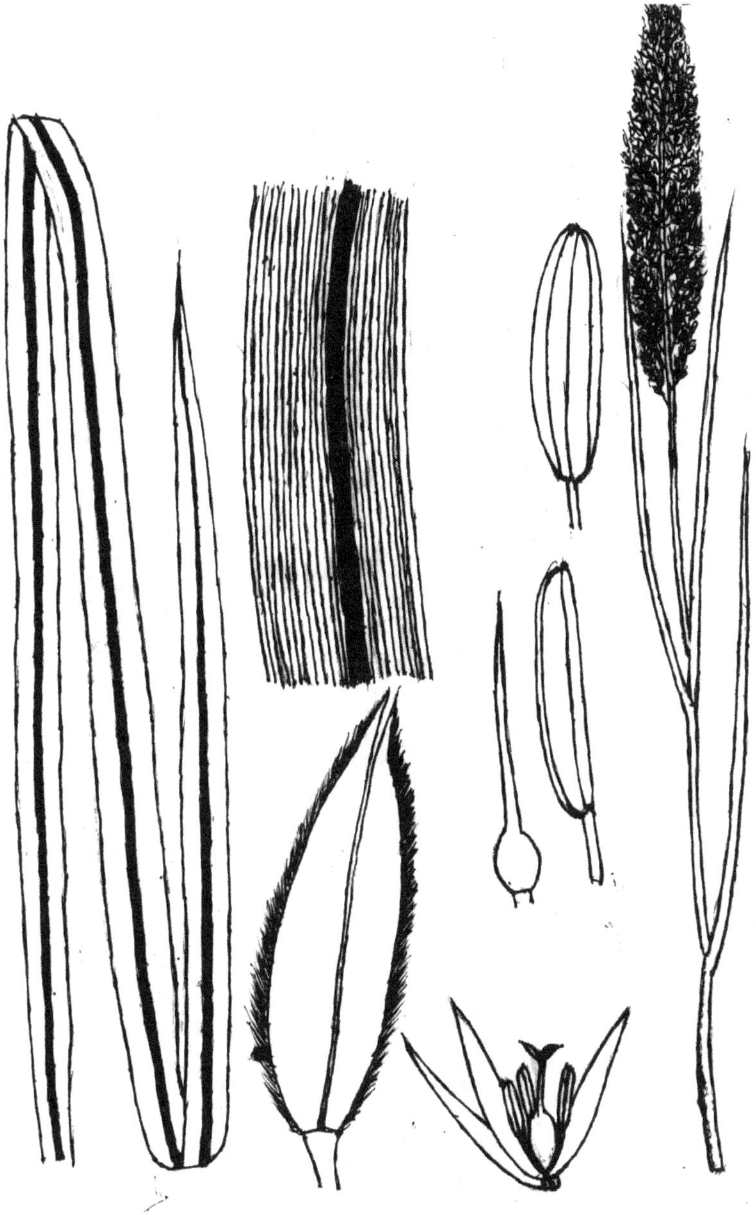

Plate 39 *Arundo donax. L var. lagopoides* Fig. 1 = A leaf; 2 = A portion of leaf; 3 = A branch with in floresens; 4 = stamen, face view and side view, 5 = A carpel; 6 = A flower; 7 = a seed.

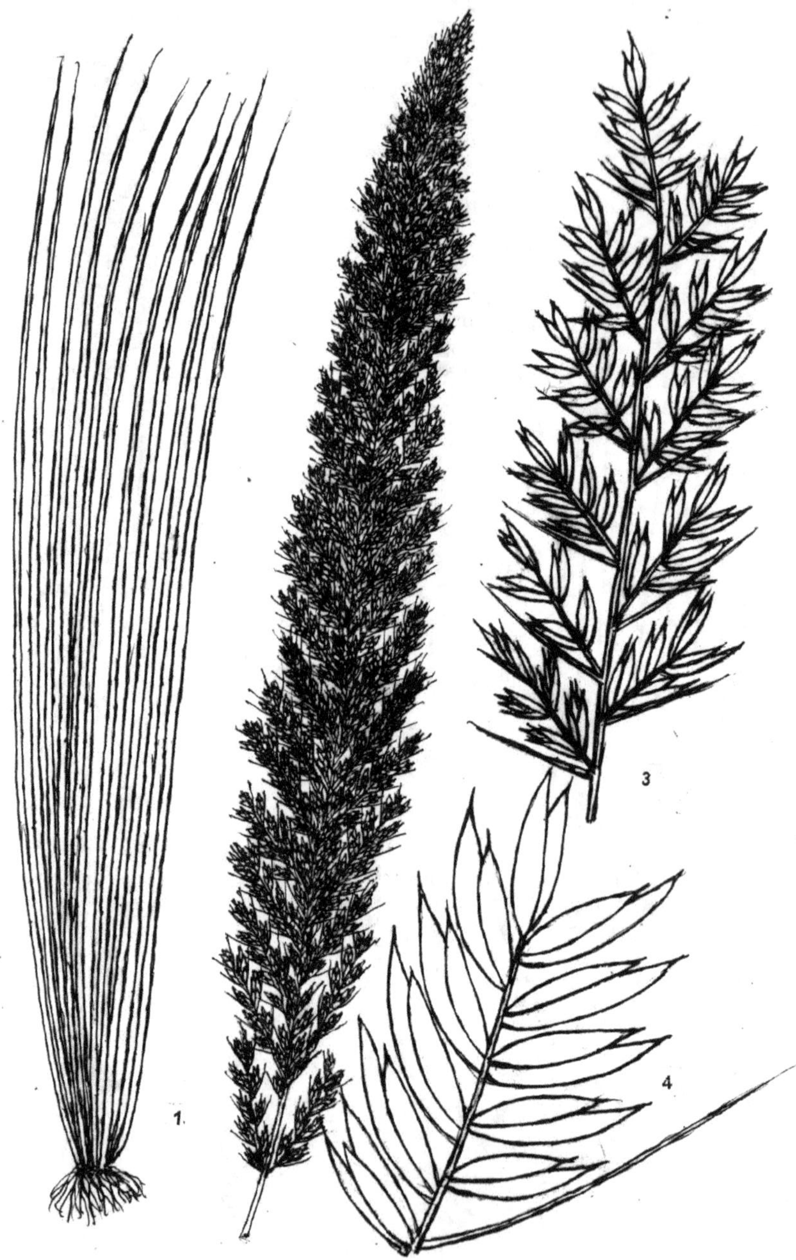

Plate 39a   *Arundo donax. L* var. *lagopoides* Fig. 1 = A plant; 2 An inflorescences; 3 : A spike; 4 = a spikelet.

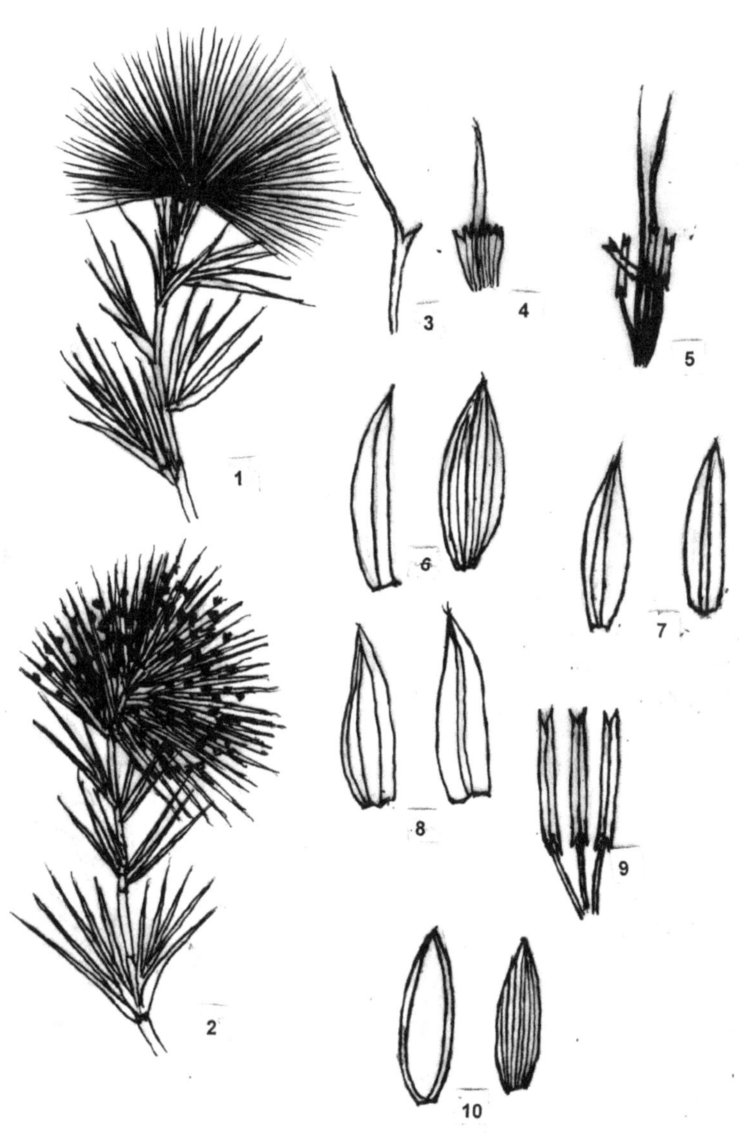

Plate 40 *Spinifix littoreus* Fig. 1 = A portion of male plant; 2 = A portion of female plant. 3 = Ligule; 4 = Bract; 5 = stamens & pistillodes; 6 = Empty glumes; 7 and 8 = Glumes; 9 = Stamens; 10 = Fruit and Seeds.

TAXONOMY

Identification key to the genera*

Ferns

A large coarse fern with pinnate leaves, arising in a cluster from the base or root stock.

- Acrostichum

Trees or shrubs showing vivipary

Not bearing any specialized root or pneumatophore; radicle about 3 to 4 cm long, curved

- Aegiceras

Bearing specialized root or pneumatophore, negatively geotrophic, pencil shaped; Leaves and branches opposite, leaves close grey, tomentose on ventral surface, hairless on dorsal surface; flowers in composed stalked heads

- Avicennia

Tall trees; knee roots thick, leaves elliptical; sepals and petals eight to fourteen; radicle cylindrical, narrowed downward, faintly ribbed

- Bruguiera

Small trees or shrubs, knee roots thin, arched; leaves obovate; sepals and petals five, radicle deeply grooved and angled, thick towards the bottom

- Ceriops

Stilt roots from base of trunk, prop roots hanging from branches; leaves more or less elliptic, acute; sepals and petals four; tip of radicle not so sharp.

- Rhizophora

Leaves more or less oblong, obovate, rounded or notched; sepals and petals five to six, tip of the radicle sharp and narrowed.

- Kandelia

Trees or shrubs not showing vivipary

Pneumatophores or special type of plank roots, pneumatophores conical, stout, woody, negatively geotrophic; fruits not vivipary, globose

- Sonneratia

Roots simple, emerging above the soil surface and curving down below again forming arch; leaves simple, narrowly elliptic with a rounded notched apex; fruits not vivipary

- Lumnitzera

Leaves paripinnate with one to three pairs of opposite leaflets; flowers in lax axillary panicles, showy

- Xylocarpus

Leaves imparipinnate, leaflets alternate; flowers small in axillary cyme

- Dalbergia

Leaves imparipinnate, leaflets opposite; flowers in fascicles, rarely solitary

- Derris

Leaves without any interpetiolar stipule, elliptic, margin spiny, toothed or entire, flowers in spikes

- Acanthus

Plants producing milky sap; leaves and branches alternate; leaves elliptic, glabrous with crenate margin; flowers dioecious, in small sessile clusters.

- Excoecaria

Plants usually succulent herbs

Succulent herbs; leaves 0; branches jointed; flowers hermaphrodite in cylindrical spikes, sunk in decussately opposite cavities of the internodes

- Salicornia

Succulent herbs; leaves terete; flowers hermaphrodite, axillary

- Suaeda

Succulent herbs; leaves opposite, entire; flowers axillary, sessile or peduncled, solitary or clustered

- Sesuvium

Plants non-succulent grasses

Non-succulent herbs; leaves much-branched, distichous, coriaceous, convolute, pungent; flowers in terminal crowded heads

- Aeluropus

*Adopted mainly from Dagar et al. (1991) and modified by Kannan et al (1999). Wherever necessary the plants not available from Tamil Nadu are excluded by the author.

Identification key to the species

Family : *Pteridaceae*

Genus : *Acrostichum*

Plants large, apex of leaflet mucronate

- *Acrostichum aureum* L.

Plants small, apex of leaflet attenuate

- *A. speciosum* Willd.

Family: *Fabaceae*

Genus: *Dalbergia*

Small erect branchlets ending in spines; leaves compound, leaflets alternate; inflorescence in axillary racemes; flowers minute; bracteoles usually persistent; pod thickened, reniform, falcate

- *Dalbergia spinosa* Roxb.

Genus: *Derris*

Leaflets many, opposite, elliptic oblong, oblong-lanceolate or the end one obovate or oblanceolate, coriaceous, shiny above; pods flat, thin, winged down one or both sides

- Derris scandens (Roxb.) Benth.

Stem dark purple, smooth, young parts pubescent; leaflets opposite, elliptic oblong, subcoriaceous; pods tapering at both ends, winged on the upper suture

- D. trifoliata

Family : *Rhizophoraceae*

Genus : *Rhizophora*

Inflorescence in leaf scar axils; flowers sessile, usually in twos, petals glabrous

- R. apiculata Bl.

Inflorescence two to eight flowered, stamens mostly eight; free part of ovary emerging beyond disc in anthers

- R. mucronata Lamk.

Genus : *Ceriops*

Shrubs or small trees, 2-3 m tall; flowers in condensed axillary cymes; hypocotyls sharply ridged and warty towards the apex

- Ceriops decandra (Griff.) Ding Hou

Genus : *Kandelia*

Tree, 5-6 m tall, bark smooth and reddish; leaves simple, glabrous, elliptic oblong, obtuse, narrowed to the base, dark green, shiny above; flowers pretty in axillary cymes

- Kandelia candel (L.) Druce.

Genns : *Bruguiera*

A medium sized to tall buttressed tree; bark smooth, grey with few lenticel; knee-like roots formed frequently; leaves 7-10 cm long, oblanceolate, rarely elliptic, acute, base cuneate; flowers about 1 cm long, white

- Bruguiera cylindrica (L.) Bl.

A straight stemmed, buttressed tree; bark dark coloured, cracked lengthwise and cross-wise; knee roots prominent; leaves crowded at the ends of branchlets, 10-16 cm long, broadly lanceolate, finely pointed, pale beneath

- B. gymnorrhiza (L.) Lamk.

Family : *Aizoaceae*

Genus : *Sesuvium*

A fleshly, prostrate and stout-stemmed rooting herb; leaves linear-oblanceolate or spathulate; sepals purplish; seeds black, shiny

- Sesuvium portulacastrum L.

Family: *Myrsinaceae*

Genus: *Aegiceras*

Leaves alternate, coriaceous and glabrous, elliptic, 4-7cm long, apex round to obtuse, margins recurved, lateral nerves about 10 pairs, inconspicuous; flowers white in terminal or axillary umbels.

Aegiceras corniculatum (L.) Blanco.

Family : *Acanthaceae*

Genus : *Acanthus*

Leaves spinous along the margins, ending in a sharp spine; flowers 3.5 to 4 cm long, bright blue or violet; bracteoles persistent

- Acanthus ilicifolius L.

Family: *Avicenniaceae*

Genus: *Avicennia*

Leaves elliptic-oblong or ovate lanceolate to elliptic lanceolate, glaucous, dull grey or whitish, tomentose beneath, glabrous above, obtuse or acute; flowers orange coloured, in 10 to 32 flowered heads

- Avicennia marina (Forsk.) Vierh.

Leaves ovate or lanceolate, acute at apex, never glaucous, cuneate at base, whitish underneath and shiny above, coriaceous, glabrous, smooth, obtuse; Flower heads capitate, globose and have rancid or fetid smell

- A. officinalis L.

Leaves lanceolate, very acute or acuminate, glabrous and shiny above, closely whitish pubescent beneath, attenuate and decurrent at base; Flowers spicate, distal, 10-30 per unit

- A. alba Blume.

Family : *Chenopodiaceae*

Genus : *Arthrocnemum*

A prostrate, glaucous-green, fleshy undershrub; branches numerous; utricle crustaceous; seeds compressed, albuminous

- Arthrocnemum indicum (Willd.) Moq.

Genus : *Salicornia*

An erect undershrub; flower-spikes very slender

- Salicornia brachiata Roxb.

Genus: *Suaeda*

Erect or ascending herb; stem glabrous, woody, much branched, reddish purple

- Suaeda maritima (L.) Dumort.

Erect decumbent herb; stems 50-100cm tall, glabrous, much branched, woody at the base, rooting at node.

- S. monoica Forsk.

Erect or ascending herb, 40-80 cm tall diffusely branched, stems much branched from woody base, smooth and yellowish, reddish

- S. nudiflora (Willd.) Moq.

Family: *Euphorbiaceae*

Genus: *Excoecaria*

Leaves alternate, ovate-elliptic or orbicular, shortly acuminate, entire or sinuate-crenate, glabrous to 7cm long, 5cm broad; flowers in catkin spikes.

- Excoecaria agallocha L.

Family : *Poaceae (Gramineae)*

Genus : *Aleuropus*

Roots long, wiry; culms tufted; leaves sometimes flat; inflorescence a globose head of dense spikelets; flowers elliptic oblong, in clusters of 4-8

- Aleuropus lagopoides (L.) Trin ex Thw.

REFERENCES

1. Blasco, F. (1975). Les Mangroves De L'inde (The Mangroves of India). Institute of Francais De Pondicherry, Trevaus de la section scientifique et Technique, Tome XIV, Fascicule **1**: 1-175.

2. Dagar, J.C. (1982). Some ecological aspects of mangrove vegetation of the Andaman and Nicobar Islands in India. *Sylvatrop Philipp. For. Res. J.*, **7:** 177-216.

3. Dagar, J.C. (1987). Mangrove vegetation, its structure, ecology, management and importance with special reference to Andaman and Nicobar Islands. In: *Proc. Symp. Management of Coastal Ecosystems and Oceanic Resources of the Andamans*, 17-18 July, 1987. Andaman Science Association (CART), Port Blair, 8-23.

4. Dagar, J.C, A.D. Mongia and A.K Bandyopadhyay. (1991). *Mangroves of Andaman and Nicobar Islands*. Oxford & IBH Publishing Co. Pvt. Ltd., New Delhi, 166pp.

5. FAO (1988). World wide compendium of mangrove - associated aquatic species of economic importance. *FAO Fish*. No. 814(FIRI/C814), 236pp.

6. Gopal, B. and K. Krishnamurthy. (1993). Wetlands of South Asia. In: *Wetlands of the world I: Handbook of Vegetation Science. Inventory, Ecology and Management* (eds.), Whigham, D.F., D. Dykyijova and S. Hejny, 345-414.

7. Kapetsky. J.M. (1985). Mangroves, Fisheries and Aquaculture. *FAO Fish Report* 338 (Suppl: 17-38).

8. Krishnamurthy, K. (1975). Socio - economic aspects of Indian mangroves. *Proc. Intemat. Symp. Biol, and Managem. Mangroves*, Univ. Florida, Gainesville, (Eds.) G. Walsh, S. Snedaker and H. Teas, Vol. **II:** 729-731.

9. Krishnamurthy, K., A. Choudhary and A.G. Untawale. (1987). Status report on mangroves on India, Ministry of Environmental and Forests, Government of India, New Delhi, 150pp.

10. Rahman Abdul, A. (1987). Ecosystem studies and management in the coastal left of Cauvery Delta at Muthupet, Tamilnadu. *Proc. Nat. Seminar on Estuarine Management.* pp. 168-171. Trivandrum June 4-5, 1987.

11. Rahman Abdul, A. (1987). An Environment investigation of Muthupet saline swamp. UGC. Research project report, 1984-1987.

12. Rahman Abdul, A and Ponnusamy P.K. (1989). Protection and Management of Mangrove estuarine wetlands in coastal belt of Cauvery delta. *Sym. Natural Resources, their conservation.* AVC College, Mannampandal, Tamilnadu. Abst, 11.

13. Rahman Abdul, A. (1990). *"The Fertility of Mangrove Forests"* in Tamil, Gurukulam printers, Vedaranyam, India.

14. Roy Chaudury, K.C. (1964). Prospects of better utilization of Goran and mangrove barks from Sunderbans for the tannin industry. *Symp. On role of forestry in the Development of Sunderbans* 24 - Parganas, India, 183-193.

15. Babcock, E.B. (1947). The genus *Crepis* I and II. *Univ. Calif. Publ. Bot.* 21 and 22:1030.

16. Fagerlind, F. (1937). Embryologische zytologische and bestaubimgs experimentalle Studien in der Familie Rubiaceae nubst Bemerkungen tiber einige polyploiditatsprobleme. *Acta Hort. Bergiani.* 11(9): 195-470.

17. Fedorov, A.N.A. (1964). *The chromosome number of Flowering Plants.* Otto Keoltz Science Publishers, n-624, Koenigstein, West Germany.

18. Gamble, J.S. (1957). *Flora of the Presidency of Madras.* Vols. 1,2 and 3. B.S.I Publication, Calcutta.

19. Homeyer, H. (1932). Zur Zytologie der Rubiaceen, *Planta,* **18:** 640.

20. King, J.R.and Bamford. (1937). The chromosome number of *Ipomoea* and related genera. *J. Heredity* **28**(8): 279-282.

21. Koul, A.K. and Singh Chauhan, B.P. (1962). Chromosome number of *Salvadora Persica* L. *Current Sci.* **31** (11):476.

22. Marimuthu, K.M. and Subramanian, M.K. (1960). A haematoxylin squash method for the root tips of *Dolichos lablab* L. *Curr. Sci.* **29**; 482-483.

23. Narayanan, C.R. (1951a). Somatic chromosomes in the Acanthaceae. *J.Madras. Univ.* **2IB**: 140-141.

24. Mathew, K.M. (1991). *An excursion flora of central Tamil Nadu, India,* Oxford & IBH Publishing Co. Pvt. Ltd. Delhi, Bombay, Calcutta.

25. Narayana, C.R. (1951b). Nucleolar behaviour and chromosomal aberrations in mitosis of *Acanthus ilicifolius* and *Asystasia coromandeliana*. *Indian J. Genet. Pl Breed*. **2**(2):205-210.

26. Patil, R.J. (1958). Chromosome numbers of some Dicotyledons. *Current Sci*. **27**(4): 140-141.

27. Raghavan, T.S. and Srinivasan, V.K. (1940). Studies in the Indian Aizoaceae. *Annals of Botany*: **2**(15):651-662.

28. Sharma, A.K. and Bhattacharrya, N.K. (1956). Cytogenetics of some members of Portulaceae and related families, *Caryologia* **8**(2): 257-274.

29. Sharma and Datta, P.C. (1958). Cytological investigations on the genus *Ipomoea* and its importance in the study of phylogeny. *Nucleus*, **1**: 89-122.

30. Sugiura, T. (1938). A list of chromosome numbers in Angiosperm plants V. *Proc. Imp. Acad. Tokyo*, **14**(10) : 391-392.

31. Tobgy, H.A. (1943). A Cytological study of *Crepis filiginosa, C.neglecta* and their Fl hybrid and its bearing on the mechanism of phylogenetic reduction in chromosome number. *J. Genetics* **45** (1):67- 111.

32. Sampath Kumar, R. (1971). "*Flora of Annamalainagar and its Neighbourhood*", Annamalai University Publications, Annamalainagar. Pages 1 to 23.

33. Subramanian, D. (1988). Cytological studies of some Mangrove Flora of Tamilnadu". *Cytologia*. **53** : 87-92.

34. FAO. (1994). Mangrove forest management guidelines. FAO Forestry Paper No. 117. *Rome*. 319 pp.

35. Fisher, P. & Spalding, M.D. (1993). *Protected areas with Mangrove habitat*. Cambridge, UK, World Conservation Monitoring Centre, (draft report).

36. IUCN. (1983). Global status of mangrove ecosystems. Commission on Ecology Papers No. 3. P. Saenger, EJ. Hegerl and J.D.S Davie, eds. Gland, Switzerland, International Union for Conservation of Nature and Natural Resources.

37. Spalding, MJX, Bfauco, F. & Field, C.D., eds. (1997). *World Mangrove Atlas*. Okinawa, Japan, The International Society for Mangrove Ecosystems.

38. Swaminathan, M.S. and Sanjay Deshmukh. (1995). Genetic engineering and adaption to climatic changes. Establishment of a genetic resources center for identifying and conserving candidate genes for use in the development of transgenic plants. Final report, *CRSARD*, Madras, India. 1995.

EXPLANATIONS OF PLATES AND FIGURES

Plate – 1:   *Acrostichum aureum* – A plant

Plate – 2:   *Acrostichum speciosum* - A plant

Plate - 3:

Figs. 1 to 3 = ***Derris scandens***

Fig. 1 = A twig with fruits; 2 = An inflorescens 3 = A flower enlarged

Figs. 4 & 5 = ***Derris trifoliata***

Fig. 4 = A twig with flowers

Fig. 5 = Fruit cluster

Plate – 4:   ***Azima tetracantha***

Fig. 1 = A twig; 2 = A twig with fruits; 3 = Ovary & Stamens; 4 = Ovary; 5 = L.S. Ovary; 6 = T.S. Ovary 7 = Stamens; 8 = Calyx 9 = Corolla; 10 – Bract and Bracteoles; 11 = Staminode; 12 = Fruit

Plate – 5:   ***Salvadora Persica***

Fig. 1 = A twig with inflorescenses; 2 = A flower = 3 = Calyx; 4 = Corolla; 5 = Petals with stamens; 6 = Stamens; 7 = T.S. and L.S. Ovary; 8 = A Fruit

Plate – 6:   ***Lumnitzera racemosa***

Fig. 1 = A twig with flowers; 2 = Petal; 3 = Stamens; 4 = Flower; 5 = Stamens; 6 = Ovary L.S; 7 = Ovary T.S.; 8 = Floral diagram.

Plate – 7:   ***Sonneratia apetala***

Fig. 1 = A twig with flowers; 2 = Flower; 3 = L.S. Flower; 4 = Sepal; 5 = Stamen; 6 = Ovary; 7 = T.S. Ovary; 8 = L.S. ovary; 9 = A fruit; 10 = T.S. Stem; 11 = Floral diagram.

Plate – 8:   *Bruigiera cylindrica*

Fig. 1 = A twig with flowers; 2 = Flower; 3 = Sepal; 4 = Petal; 5 = Stamen; 6 = T.S. and L.S. Ovary; 7 = Fruit; 8 = Floral diagram.

Plate – 9:   *Bruigiera gymnorrhiza*

Fig. 1 = A twig with flowers; 2 = A flower; 3 = L.S. Flower; 4 = T.S. Stem; 5 = Floral diagram; 6 = Sepal, Petal and Stamen; 7 = T.S. Ovary; 8 = L.S. ovary; 9 = A fruit.

Plate – 10:

Figs. 1 & 2 = **Ceriops tagal;** Fig. 1 = A twig; 2 = A fruit; Figs. 3 to 5 = **Ceriops candolleana;** Fig. 3 = A twig; 4 = Young fruits; 5 = A mature fruit.

Plate – 11:   *Ceriops decandra*

Fig. 1 = A twig; 2 = A flower; 3 = bract; 4 = Sepal; 5 = Petal; 6 = Stamen; 7 = Ovary; 8 = T.S. and L.S. Ovary; 9 = Young Fruit; 10 = Fruits; 11 = Floral diagram.

Plate – 12:   *Rhizophora mucronata*

Fig. 1 = A twig; 2 = Flower; 3 = Calyx; 4 = Petal; 5 = Sepal; 6 = Pistil; 7 = T.S. and L.S. Ovary; 8 = Stamens; 9 = A fruit.

Plate – 13:   *Rhizophora species*

Plate – 14:   *Rhizophora apiculata*

Fig. 1 = A twig; 2 = A flower; 3 = Bracteoles; 4 = Sepal; 5 = Petal; 6 = Stamen; 7 = T.S. and L.S. Ovary; 8 = Stamen; 9 = Fruit.

Plate – 15:   *Rhizophora* – a peculiar population

Plate – 16:   *Rhizophora hybrid* (R. apiculata x R. mucronata)

Plate – 17:   *Kandelia candel*

Fig. 1 = A twig; 2 = A leaf; 3 = Flowers; 4 = Young fruit; 5 = Fruit.

Plate – 18:   *Floral diagrams*

Plate – 19:   *Sesuvium portulacastrum*

Fig. 1 = A plant; 2 = A flower; 3 = A flower less petals; 4 = Ovary; 5 = T.S. Ovary; 6 = L.S. Ovary; 7 = A. Fruit; 8 = A seed.

Plate – 20: ***Aegiceros corniculatum***

Fig. 1 = A twig; 2 = A young fruit cluster; 3 = Flower cluster; 4 = Flower; 5 = Stamen; 6 = Petals and Stamens; 7 = Ovary; 8 = T.S and L.S. ovary; 9 = Fruit Clusters.

Plate – 21: ***Aegiceros majus***

Fig. 1 = A plant; from kille 2 = Fruit cluster; 3 = A plant from Cuddalore.

Plate – 22: ***Ipomea biloba***

Fig. 1 = A twig; 2 = T.S. stem, 3 = A flower; 4 = T.S. Ovary, 5 = Stamens; 6 = Fruit; 7 = Floral diagram.

Plate – 23: ***Acanthus ilicifolius***

Fig. 1 = A twig; 2. flower, side view; 3 = Flower, face view, 4 = Fruit cluster; 5 = Ovary; 6 = Sepal, petals; 7 = Stamens; 8 = A fruit.

Plate – 24: ***Avicennia alba***

Plate – 25: ***Avicennia marina***

Fig. 1 = A twig; 2 = A flower; 3 = Sepal; 4 & 5 = Petals; 6 = Stamen; 7 = T.S. Ovary; 8 = L.S. Ovary; 9 = Fruit cluster.

Plate – 26: ***Avicennia officinalis***

Fig. 1 = A twig; 2 = Flower face view; 3 = Flower, side view; 4 = L.S. Flower; 5 = Sepals & Petal; 6 = Stamen; 7 = T.S. Ovary; 8 = Fruit Clusters; 9 = Floral diagram.

Plate – 27: ***Arthrocnemum indicum***

Fig. 1 = Vegetative shoot; 2 & 3 = Flowering shoots; 4 = Flowers; 5 = Ovary and Stamen; 6 = T.S. Ovary; 7 = Stamens; 8 = Condensed shoot.

Plate – 28: ***Suaeda monoica***

Fig. 1 = Vegetative shoot; 2 = Flowering shoot; 3 = A flower; 4 = Bracts & Bracteoles; 5 = Petals & stamens; 6 = Stamens; 7 = Ovary; 8 = T.S. and L.S. Ovary.

Plate – 29: ***Suaeda nudiflora***

Fig. 1 = Vegetative shoot; 2 = Flowering shoot; 3 = Flowers; 4 = Flower; 5 = Petals & Stamens; 6 = Bracts & Bracleoles; 7 = A petal and a Stamen; 8 = Ovary; 9 = T.S. and L.S. Ovary.

Plate – 30:   *Excoecaria agallocha*

Fig. 1 = A twig; 2 = A fruiting twig; 3 = A female flower; 4 = Stamens; 5 = Carpel; 6 = T.S. and L.S. Ovary; 7 = Male inflorecens; 8 = Stamens; 9 = Fruit; 10 = Floral diagrsm, male and female.

Plate – 31:

Figs. 1 to 6 = **Cyperus arenarius;** Fig. 1 = A plant; 2 = Spike; 3 = Bract & sepals; 4 = Rachilla; 5 = A flower; 6 = Fruit; Figs. 7 to 12 = **Fimbristylis cymosa;** Fig. 7 = A plant; 8 = Spike; 9 = Glumes; 10 = Rachilla; 11 = A Flower; 12 = Fruit.

Plate – 32:

Figs. 1 to 7 = **Fimbristylis lagopoides**

Fig. 1 = A plant; 2 = Spikes; 3 = Spike enlarged; 4 = Flowers; 5 = A flower; 6 = Ovary; 7 = Glumes;' Figs. 8 to 13 = **Cyperus portonovensis;** Fig. 8 = A plant; 9 = Spikes; 10 = Spike enlarged; 11 = Spikelet; 12 = A flower; 13 = Glumes; 14 = Fruit.

Plate – 33:   *Fimbristylis spathacea*

Fig. 1 = Plant; 2 = Spikes; 3 – Few spikes enlarged; 4 = Spikelet; 5 = Glume; 6 = Ovary; 7 = Stamen.

Plate – 34:

Figs. 1 to 6 = Fimbristylis polytrichoides

Fig. 1 = Plant; 2 = Spike enlarged; 3 = Flower, 4 = Bract sepal & petal; 5 = Rachilla; 6 = Ovary; Figs. 7 to 10 = Portiresia coarctata; Fig. 7 = Plant; 8 = Leaf; 9 = Fruit; 10 = Seed.

Plate – 35:   *Aeluropus lagopoides*

Fig. 1 = Plant; 2 = Glume; 3 = Lower and Upper glumes; 4 = Flowering glumes; 5 = Limb; 6 = Stamens; 7 = Stamen; 8 = Ovary.

Plate – 36:   *Coelachryum lagopoides*

Fig. 1 = Plant; 2 = Spike cluster; 3 = Spike; 4 = Upper glumes; 5 = Flowering glumes; 6 = Lemma; 7 = palaea; 8 = Flower; 9 = Ovary; 10 – Stamens; 11 = Caryopsis; 12 = Limb.

Plate – 37:

Fig. 1 = ***Cynodon dactylon***

Long prostrate twining and crawling branch.

Fig. 2 to 7 = ***Dactyloctenium aegypticum***

Fig. 2 = Long prostrate crawling branch; 3 = Plant tip with spikes; 4 = Spikes enlarged; 5 = A spike; 6 = Glumes; 7 = Stamens; 8 = Ovary.

Plate – 38

Figs. 1 to 5 = ***Sporobolus diander***

Fig. 1 = A plant; 2 = Ligule; 3 = Upper and Lower Glumes; 4 = Lemma; 5 = Flower and Palea

Figs. 6 to 8 = ***Zoisia matrella***

Fig. 6 = A plant; 7 = A flower; 8 = A fruit

Plate – 39   ***Arundo donax. L*** var. ***lagopoides***

Fig. 1 = A leaf; 2 = A portion of leaf; 3 = A branch with in floresens; 4 = stamen, face view and side view, 5 = A carpel; 6 = A flower; 7 = a seed.

Plate – 39a   ***Arundo donax. L*** var. ***lagopoides***

Fig. 1 = A plant; 2 An inflorescences; 3 : A spike; 4 = a spikelet.

Plate – 40:   ***Spinifix littoreus***

Fig. 1 = A portion of male plant; 2 = A portion of female plant.

3 = Ligule; 4 = Bract; 5 = stamens & pistillodes; 6 = Empty glumes; 7 and 8 = Glumes; 9 = Stamens; 10 = Fruit and Seeds.

PART – II

MARINE ANGIOSPERMS OF TAMIL NADU

INTRODUCTION

The Marine Angiosperms or Sea grasses come under two Monocotyledonous families, Potamogetonaceae and Hydrocharitaceae. Enhalus, Halophila and Thalassia come under Hydrocharitaceae. Cymodocea, Holodule and Syringodium come under Potamogetonaceae. Totally 13 species, one species with 2 subspecies, come under the above said 6 genera, in Eastern coast of Tamil Nadu. K. Ramamurthy et al., (1992) have described the ecological, morphological and taxonomical aspects of these 13 species in their book "Seagrasses of Coromandel coast, India". Therefore, the author is providing here the cytological and more important morphological features of these taxa with illustrations. Recently a few scientists have done some experiments on how the plant extracts of sea grasses can be useful for crop improvement. R. Ragupathy Raja, Kannan et al., (2010) have made phytochemical analysis of sea grasses from Mandapam coast, Tamil Nadu. They studied the presence or absence of alkaloids, coumarin, flavonoids, glycosides, phenol, protein and aminoacids, quinone, saponine, sterols, sugar, tannin and terpinoids in Halophila stipulacea, Cymodocea serrulata and Holodule pinifolia. They have also analysed the % of biochemicals, that is protein, carbohydrate and lipid and amount of pigments (mg/g) that is chlorophyll a, chlorophyll b and total chlorophyll and carotenoids in these three plants.

The observations on the occurrence of marine angiosperms in Tamil Nadu shows that the more we go to Southern districts the more will be the availability of these plants. This shows that the rocky substratum and coral reef beds are more in Southern parts near seashore, and so more amount of marine angiosperms are present. Enhalus acoroides and Syringodium isoetigolium are present only in Ramanathapuram and Kanyakumari districts with reproductive parts. "Seagrasses of coromandel coast India" by Ramamurthy et al., (1992) is an illustrated book which is useful to understand all the 14 species of marine angiosperms. The economic importance of these plants has not been fully understood. It is said they serve as sediment traps besides stabilizing the bottom sediments; thereby improving the water quality. They provide with food and shelter for diverse organisms, particularly fishes. Some of the research papers published recently describe repeatedly the general information of seagrasses. It is a group of flowering plants that are totally immersed under salt water and are established permanently in that unusual environment. Therefore, they should have some special architecture to overcome these adverse conditions. Therefore, it is necessary to work out the inherent potentialities and chemical compositions of these seagrasses which may be useful for human beings in one or other way. Repeating the same general information on seagrasses is a mere waste.

The author has studied the cytological characters of two plants, *Halophila ovalis* and *Najus spinosa* var. *minor* in 2000. Of these the latter plant was present inside the salt water in an old cement tank along with salt water form of *Chara zeylanica*. This plant is alive inside the salt water. But in Gamble's "Flora of the presidency of Madras" it was stated that it was present in Madras. No information about whether it is aquatic or terrestrial is available. The author fixed the root tips at spot of collection and did cytology successfully. Therefore it is sure that it is a sea water plant or it is capable of living in sea water. The author is trying to collect them from seashores or lagoons, estuaries or backwater canals. Final decisions will be made after confirming its presence inside salt water or seashores to add this plant as aquatic angiosperms.

Among the 9 species of marine angiosperms, in which cytological studies have been made so far, the author has made such studies in 8 species. Janaki Ammal (1945) has studied the cytological characters of *Enhalus acoroides*. Deviant report of chromosome number has been made in *Halophila ovalis* by the author (2n = 16) as against the previous report of Harada (1951, 1955) that is 2n=18. In all the other 7 species, cytological studies have been made by the author for the first time in 2010.

Ramamurthy *et al.*, (1992) have given a chronological list of scientists who have collected directly the live plants under sea-water in a particular place or the broken pieces of plants washed ashore in some places. This data is important for those interested in cytological studies. Washed ashore plants and their broken parts will not be useful for cytological studies as they are almost dead without active cell division of any part. To collect the live plants directly under sea-water, we have to go to a particular seashore and stay that night and early morning as we have to reach the seashore by 7 to 8 a.m. and fix the root tips or shoot tips or flower buds. At this time, there may be low tides and it is easier to collect plant parts, as the plants are almost exposed. Only in the morning time, there may be active cell divisions in all the parts of the plant and the nucleus will transform into bivalents or chromosomes. At this stage, the root tip, leaf tips and flower buds are arrested without further growth and so the exact condition is maintained for a few days by fixing them in a fixative 1:3 acetic alcohol.

A large number of scientists have given ecological maps of live centers of marine angiosperms and it will be useful to collect a particular plant in a particular place.

Arumugam *et al.,* (2010) have studied the antimicrobial potential of some seagrasses, namely *Halophila stipulacea*, *Cymodocea serrulata* and *Halodule pinifolia* against some phytopathogens. Balasubramanian *et al.,* (2000) screened some seagrasses for antibacterial activity against bacterial pathogens.

OBSERVATIONS AND DISCUSSIONS

The author provides below two topics 1. Cytological studies and 2. Taxonomical studies. Even though morphological, ecological and taxonomical studies have been made here and there, the cytological studies have not been made in India. Of the 14 species of marine angiosperms in India, only in Enhalus acoroides, a cytological study was undertaken in 1945 by Janaki Ammal. There are 13 genera and 52 species present all over the world (Ramamurthy et al., 1992). But no record of cytological studies is available upto 1974, as far as the author is aware. Cytological studies are very difficult in marine angiosperms and mangroves, because of heavy salt depositions throughout the plants. But, if we try, it is possible to get success. Cytological studies are strong evidences to interpret species relationships, origin and evolution of species, to find out whether a species is mutant or hybrid or polyploidy and to highlight the relationship of a species to its related members under study.

CYTOLOGICAL STUDIES

As already pointed out, out of 52 seagrasses of the world, only in 9 species alone, cytological studies have been made so far. The author has furnished below, the cytological studies of 8 species.

For cytological studies modified method of iron alum haematoxylin squash technique was followed (Subramanian, 2010). The prepared slide was seated with nail polish or thermacol dissolved in xylol. Important mitotic stages were drawn using the microscope and Abbe camera lucida. Measurements of chromosomes were made with ocular micrometer, the scale of which was already devised from stage micrometer. With computer application, karyotype analyses were used to prepare idiograms (Plates - 2 & 3). From the absolute chromosomes lengths of various species studied, histogram (Plate - 4) was drawn to show the comparative amount of DNA and RNA present in each and every species studied. Some of the important plates were photographed with attachment camera under oil immersion lens of the trinocular compound microscope. During this study, XX and XY sex chromosomes of the male and female plants in dioecious species were identified and shown in the karyotype analyses of *Halophila ovata, H. ovalis* and *H. ovalis* var. *ramamurthiana.*

Results

Table-1 shows the species studied, places of collections, 2n chromosome numbers, authors and years. The morphology of chromosomes reveals 6 types namely Sm^{Sat}, Sm, M, m, st and t. The number of chromosomes in each type, nature of constrictions, long arm, short arm and satellite length are presented for each species in Table-2 to 9. The absolute chromosome length, average chromosome length and total chromosome length of each species were calculated. The Karyotype analyses of the seagrass species are given in Table -10. The cytological characters of the species are furnished below.

1. *Halophila beccarii Asch.*

The plants are bisexual. The diploid chromosome number is 2n = 20 (Plate-1 and Fig. 1). Most of the chromosomes have median and terminal constrictions (Table-2 and Plate- 2). The length of the chromosomes ranges from 1.4 to 4.6 μm.

2. *Halophila decipiens Osten.f.*

The plants are bisexual. The diploid chromosome number is 2n = 22 (Plate-1 and Fig. 2). Most of the chromosomes have submedian constrictions (Plate-2 and Table-3). The length of the chromosomes ranges from 2.0 to 5.0 µm. Multipolar anaphases have been rarely observed in this species.

3. *Halophila ovata Gaud.*

Plants are dioecious and unisexual. The diploid chromosome number is 2n = 16 (Plate-1 and Figs. 3 & 4). Most of the chromosomes have submedian and median constrictions (Table-4 and Plate-2). In male plants, one X and one Y chromosomes are present and they are devoid of constrictions. In female plants, two XX chromosomes are present without constrictions. There are rarely cells with 2n = 24 chromosomes (Plate-1 and Fig. 5). The chromosomes are present during metaphase within the circular configuration. There are rarely cells with precocious movements of chromosomes during metaphase and also anaphasic bridges (Plate 1 and Figs. 6 & 7). The size of the chromosomes ranges from 1.2 to 3.2 µm.

4. *Halophila ovalis Hook.f*

The plants are dioecious and unisexual. The diploid chromosome number is 2n = 22 (Plate-1 and Figs. 8 & 9). Most of the chromosomes have submedian and subterminal constrictions (Table-5 and Plate-2 and Fig. 4). The length of the chromosomes ranges from 1.6 to 5.4 µm. In female plant, there are 2 X (sex) chromosomes and in male one X- and one Y (sex) chromosomes are present. There are rarely cells with precocious movements of chromosomes at metaphases and anaphasic bridges (Plate-1 and Figs. 10 to 12).

Table 1 Chromosome numbers in different species of seagrasses

S. No.	Species	Place of collection	Present report	Previous report (n = 2n)	Author and year (n = 2n)
1.	**Enhalus acoroides**	Mandapam	-	14	Jana kiammal (1945)
2.	**Halophila ovata**	Mandapam	16$^{\#}$	18	Harada (1951-1956)
3.	**H. ovalis**	Mandapam, Porto-Novo	22*	-	-
4.	**H. beccarii**	Porto-Novo	20*	-	-
5.	**H. decipiens**	Porto-Novo, Pichavaram, T.S. Pettai	22*		
6.	*H. ovalis* ssp. *rama murthiana*	Marak kanam Thazan guppam	24*	-	-

S. No.	Species	Place of collection	Present report	Previous report (n = 2n)	Author and year (n = 2n)
7.	*Holodule pinifolia*	Pichavaram	20*	-	-
8.	*H. uninervis*	Porto-Novo, Pichavaram	22*	-	-
9.	*H. wrightii*	Pazhaiyar	22*	-	-

*First records of diploid chromosome numbers in the respective species;

#Deviant report of chromosome number as against the earlier report

Table 2 Chromosome structures in Halophila beccarii (2n = 20)

S. No.	Types of chromosome	No. of chromosome	Chromosome length in μm		Sat/ secondary arm chromo- some	L/B ratio	Total length in μm
			Long arm	Short arm			
1.	Sm$^{Sa'}$	2	3.0	2.0	0.2	1.5	5.6
2.	Sm	2	2.2	1.6	-	1.37	4.6
3.	M	8	1.4	1.4	-	1.0	3.0
4.	St	2	1.4	0.8	-	1.75	2.4
5.	t	6	1.6	0.2	-	8.0	2.0

Total chromosome length = 61.2 mrn; Absolute chromosome length = 30.6 mrn; Average chromosome length = 3.4 mrn

Table 3 Chromosome structures in *Halophila decipiens* (2n = 22)

S. No.	Types of chromosome	No. of chromosome	Chromosome length in μm		Sat/ secondary arm chromo- some	L/B ratio	Total length in μm
			Long arm	Short arm			
1.	Sm[Sa']	2	3.0	2.4	0.2	1.25	6.0
2.	Sm	2	2.4	2.0	-	1.20	4.6
3.	Sm	2	2.6	1.4	-	1.85	4.2
4.	Sm	4	2.0	1.6	-	1.25	3.8
5.	M	4	1.7	1.7	-	-	3.6
6.	M	2	2.0	1.8	-	1.11	3.2
7.	St	2	2.0	0.4	-	5.0	2.6
8.	t	4	1.6	0.2	-	8.0	2.0

Total chromosome length = 78.8 mrn; Absolute chromosome length = 39.4 mrn; Average chromosome length = 3.58 mrn

Table 4 Chromosome structures in Halophila ovata (2n = 16)

S. No.	Types of chromosome	No. of chromosome	Chromosome length in µm		Sat/ secondary arm chromo-some	L/B ratio	Total length in µm
			Long arm	Short arm			
1.	Sm$^{Sa'}$	2	1.6	1.0	0.2	1.60	3.2
2.	Sm	4	1.4	1.2	-	1.17	2.8
3.	M	4	1.0	1.0	-	-	2.2
4.	St	4	0.8	0.2	-	4.0	1.2
5.	X (for female plants 2 "XX")	-	-	-	-	-	1.0
6.	Y (for male plants, IX and IY)	-	-	-	-	-	0.8

Total chromosome length for female = 33.2 mrn; Total chromosome length for male = 33.0 mrn; Absolute chromosome length for female = 16.6 mrn; Average chromosome length for male = 16.5mrn

Table 5 Chromosome structures in Halophila ovalis (2n = 22)

S. No.	Types of chromosome	No. of chromosome	Chromosome length in μm		Sat/ secondary arm chromo-some	L/B ratio	Total length in μm
			Long arm	Short arm			
1.	Sm[Sa']	2	2.4	2.0	0.2	1.20	5.0
2.	Sm	2	2.4	1.8	0.2	1.34	4.8
3.	Sm	2	2.4	1.8	-	1.34	4.4
4.	Sm	2	3.2	0.6	-	5.33	4.0
5.	M	2	1.5	1.5	-	-	3.2
6.	M	6	1.6	1.2	-	1.33	3.0
7.	St	2	1.6	0.6	-	2.67	2.4
8.	t	2	1.4	0.2	-	7.0	2.0

S. No.	Types of chromosome	No. of chromosome	Chromosome length in μm		Sat/ secondary arm chromo-some	L/B ratio	Total length in μm
			Long arm	Short arm			
9.	X (2X for female)	2	-	-	-	-	1.8
10.	Y (1X and 1Y for male)	1	-	-	-	-	1.6

Total chromosome length for female = 73.2 mrn; Total chromosome length for male = 73.0 mrn; Absolute chromosome length for female = 36.6 mrn; Average chromosome length for male = 36.5mrn

Table 6 Chromosome structures in Halophila ovalis var ramamurthiana (2n = 22)

S. No.	Types of chromosome	No. of chromosome	Chromosome length in μm		Sat/ secondary arm chromo-some	L/B ratio	Total length in μm
			Long arm	Short arm			
1.	Sm[Sa']	2	2.2	1.0	0.2	2.20	4.0
2.	Sm	2	2.0	1.2	-	1.67	3.8
3.	M	6	1.6	1.6	-	-	3.4
4.	St	6	2.0	0.6	-	3.33	2.8
5.	T	6	1.4	0.2	-	7.0	1.8
6.	X (2X for female)	2	-	-	-	-	1.2
7.	Y (1X and 1Y for male)	1	-	-	-	-	1.0

Total chromosome length for female = 56.0 mrn; Total chromosome length for male = 55.8 mrn; Absolute chromosome length for female = 28.0 mrn; Average chromosome length for male = 27.9 mrn

Table 7   Chromosome structures in Holodule pinifolia (2n = 20)

S. No.	Types of chromosome	No. of chromosome	Chromosome length in μm		Sat/ secondary arm chromo- some	L/B ratio	Total length in μm
			Long arm	Short arm			
1.	Sm$^{Sa'}$	2	1.6	0.6	0.2	2.67	2.8
2.	Sm	4	1.4	0.8	-	1.75	2.4
3.	Sm	2	1.2	0.8	-	1.50	2.2
4.	M	2	0.9	0.9	-	-	2.0
5.	m	2	1.0	0.6	-	1.67	1.8
6.	St	2	1.0	0.4	-	2.5	1.6
7.	t	6	1.0	0.2	-	5.0	1.4

Total chromosome length = 38.6 mrn; Absolute chromosome length = 19.3 mrn; Average chromosome length = 1.93 mrn

Table 8 Chromosome structures in Halodule uninervis (2n = 22)

S. No.	Types of chromosome	No. of chromosome	Chromosome length in μm		Sat/ secondary arm chromo-some	L/B ratio	Total length in μm
			Long arm	Short arm			
1.	Sm$^{Sa'}$	2	1.2	0.6	0.2	2.0	2.4
2.	Sm	4	1.4	0.6	-	2.33	2.2
3.	M	8	0.9	0.9	-	-	2.0
4.	m	2	1.0	0.6	-	1.67	1.8
5.	St	4	1.0	0.4	-	2.5	1.6
6.	t	2	0.6	0.2	-	3.0	1.0

Total chromosome length = 41.6 mrn; Absolute chromosome length = 20.8 mrn; Average chromosome length = 1.89 mrn

Table 9 Chromosome structures in Halodule wrightii (2n = 22)

S. No.	Types of chromosome	No. of chromosome	Chromosome length in μm		Sat/ secondary arm chromo-some	L/B ratio	Total length in μm
			Long arm	Short arm			
1.	Sm$^{Sa'}$	2	2.0	1.0	0.2	2.0	3.6
2.	Sm	2	2.2	1.0	-	2.2	3.4
3.	M	2	1.5	1.5	-	-	3.0
4.	m	4	1.6	1.0	-	1.6	2.8
5.	St	8	1.6	0.6	-	2.67	2.4
6.	t	4	1.6	0.2	-	8.0	2.0

Total chromosome length = 58.4 mm; Absolute chromosome length = 29.2 mm; Average chromosome length = 2.65 mm

Table 10 Karyotype analyses of the seagrass species

S. No.	Species	"2n" number	Sm^{Sat}	Sm	M	m	St	t	Sex chromosome
1.	**Halophila. beccarii**	20	2	2	8	-	2	6	-
2.	**H. decipiens**	22	2	8	4	2	2	4	-
3.	**H. ovata**	16	2	4	4	-	4	-	2
4.	**H. ovalis**	22	4	2	2	6	4	2	2
5.	*H. ovalis* **ssp.** *ramamurthiana*	24	2	2	6	4	6	6	2
6.	**Holodule pinifolia**	20	2	6	2	2	2	6	Sex chromosomes not studied
7.	**H. uninervis**	22	2	4	8	2	4	2	Sex chromosomes not studied
8.	**H. wrightii**	22	2	2	2	4	8	4	Sex chromosomes

The plants are dioecious and unisexual. The diploid chromosome number is 2n = 24 (Plate-1 and Figs. 13 & 14). Most of the chromosomes have submedian and subterminal constrictions (Table-6 and Plate 3). The length of chromosomes ranges from 1.0 to 4.0 um. In the female plant, there are 2 X chromosomes and in the male plant one X and one Y chromosomes.

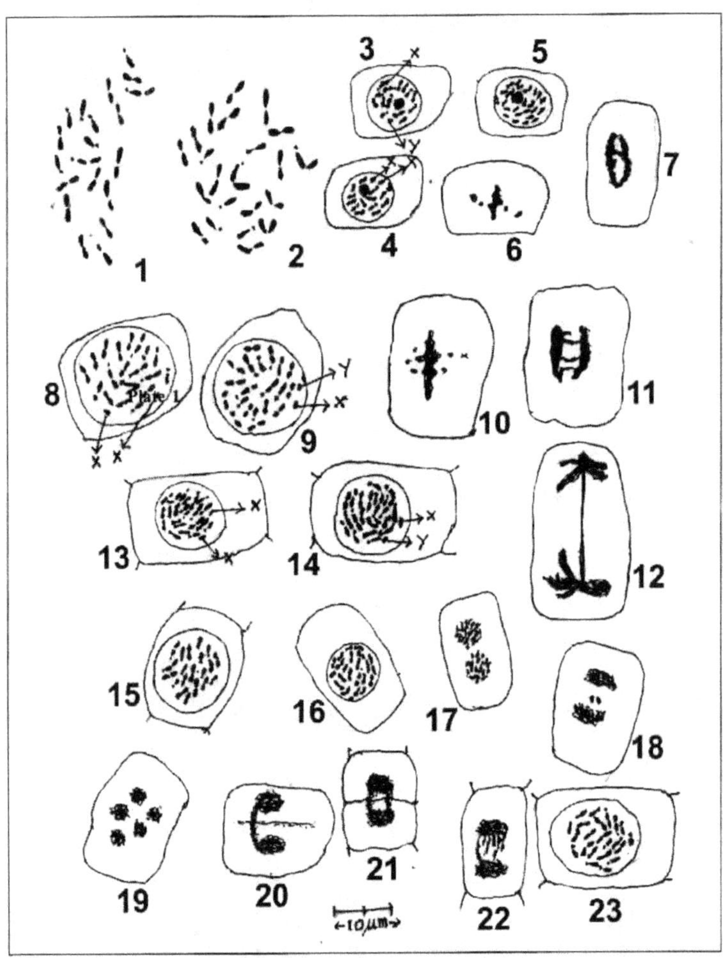

Fig. 1. - *Halophila beccarii*, 2n = 20; 2 - *H. decipiens*, 2n = 22; 3 and 4 - *H. ovata*, 2n = 16, 5 = *H. ovata*, 2n = 24; 6 and 7 - mitotic abnormalities; 8 and 9 - *H. ovalis*, 2n = 22; 10 to 12 = mitotic abnormalities; 13 and 14 - *H. ovalis* ssp. *ramamurthiana*, 2n = 24; 15 - *Holodule pinifolia*, 2n = 20; 16 - *H. uninervis*, 2n = 22; 17 to 22 - *H. uninervis*, mitotic abnormalities; 23 - *H. wrightii*, 2n = 22

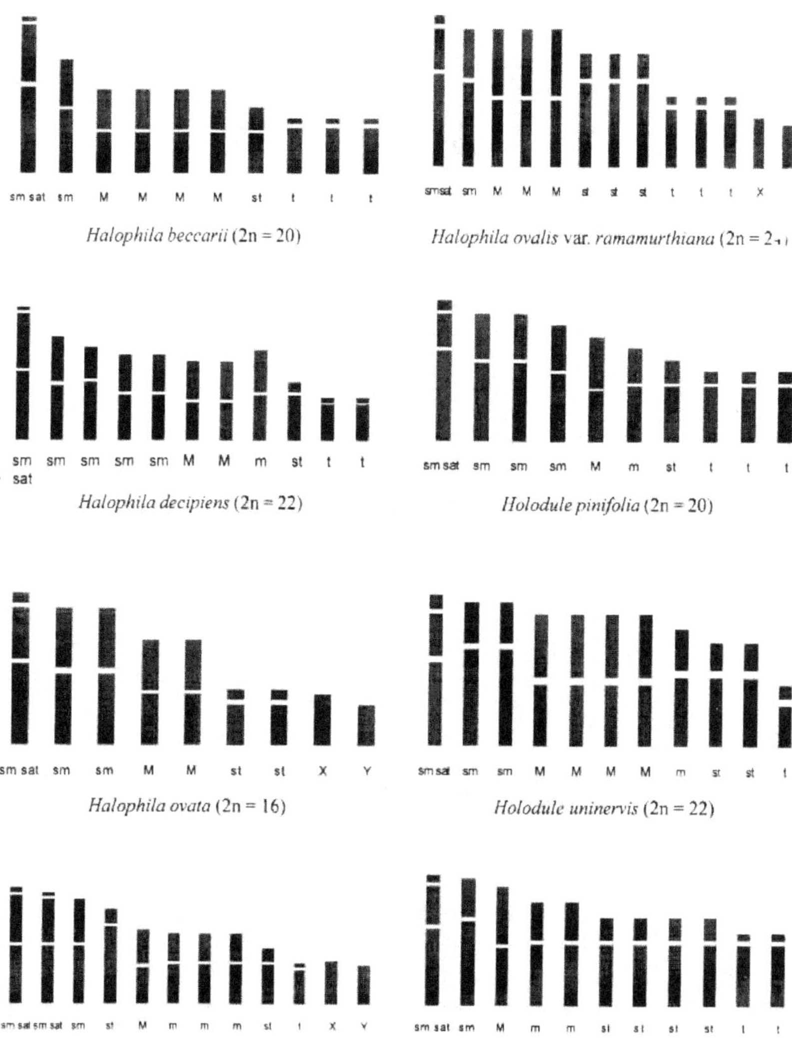

Plate 2. *Karyotype analyses of* Halophila beccarii, H. decipiiens, H. ovata *and* H.ovalis

Plate 3. *Karyotype analyses of* Halophila ovalis var. ramamurthiana, Holodule pinifolia, H. uninervis *and* H.wrightii

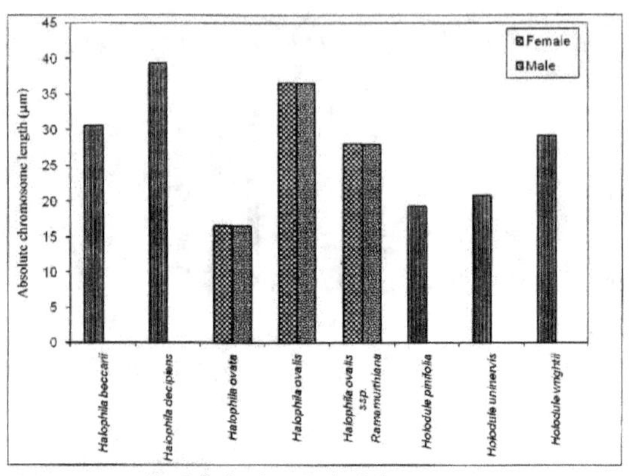

Plate 4. *Absolute chromosome length of seagrass species*

5. *Halodule pinifolia* (MM) Hartog.

The plants are dioecious and unisexual. The chromosomes are very small and unable to study the sex chromosomes of all the 3 species of Halodule. The somatic chromosome number is 2n = 20 (Plate-1 and Fig. 15). Most of the chromosomes have submedian, subterminal and terminal constrictions (Table-7 and Plate 3). The length of chromosomes ranges from 1.4 to 2.8 um. The chromosomes at metaphase, polar view are centrally placed within a circular configuration.

6. *Halodule uninervis* (Forsk) Asch.

The plants are dioecious and unisexual. The somatic chromosome number is 2n = 22 (Plate-1 and Fig. 16). Most of the chromosomes have submedian, subterminal and terminal constrictions (Table-8 and Plate 3). The length of the chromosomes ranges from 1.0 to 2.4 µm. The metaphase chromosomes are present within a circular configuration like the previous species. Besides cells with bipolar anaphase, anaphasic laggards, multipolar anaphase, broken anaphasic bridges, bridge between two nuclei of 2 daughter cells have been observed (Plate-l and Figs. 17 to 22).

7. *Halodule wrightii* Asch.

The plants are dioecious and unisexual. The diploid chromosome number is 2n = 22 (Plate-1 and Figs, 23). Most of the chromosomes have subterminal and terminal constrictions (Table 9 and Plate 3). The length of chromosomes ranges from 2.0 to 3.6 μm. The size of chromosomes, total chromosome length, absolute chromosome lengths are more in this species when compared to the 2 species of Halodule studied. The chromosomes are confined within a circular configuration inside the cell, in this species also as in the other 2 species of Halodule studied herein.

Discussion

Cytological studies have been made in 8 species of Seagrasses occurring in the Coromandal coast of Tamil Nadu. First record of somatic chromosome numbers have been made in Halophila beccarii, H. decipiens, H.ovalis, H. ovalis var. ramamurthiana, Halodule pinifolia, H. uninervis and H.wrightii. Deviant report as against the earlier record of chromosome number has been made in H. ovata (Fedorov, 1974; Virendrakumar and Subramaniam, 1982). Harada (1951) recorded 2n=18 chromosomes in Halophila ovata but the present investigation shows 2n = 16 chromosomes in this species. Halophila ovalis is a dioecious species and for the first time sex chromosomes of male and female plants have been made not only in this species but also in H. ovata, which is also a dioecious member. Halophila beccarii and H. decipiens are monoecious and male and female flowers are present in one and the same plant. These two species are available only in Vellar estuary near the seashore of Porto-novo along with H. ovalis and Halodule pinifolia. H.ovalis and H.ovata species are widespread throughout the Coromandel Coast of South India. Stebbins et' al. (1953) reported that dioecious plants are recent and more evolved enjoying more widespread area due to favourable adaptive mechanisms than their monoecious relatives which are ancient in origin but lacking adaptive mechanisms for widespread living. So, when compared to H. beccarii and H. decipiens, H. ovata, H ovalis and H. stipulacea (dioecious) are more evolved and have more widespread living conditions in the Coromandel coast of Tamil Nadu.

The Karyotype analyses of chromosomes of the species show that chromosomes are larger in *H. beccarii* and *H. decipiens,* whereas they are smaller in *H. ovata* and *H. ovalis.* According to Stebbins *et al.* (1953), the evolved species have smaller chromosomes, whereas in the primitive species the chromosomes are larger. This is true for flowering plants of the world. Hence, *H. beccarii* and *H. decipiens* are primitive plants and they find a restricted area of distribution and also possibly they are in a state of extinction. But, in general, the seagrasses studied are more evolved as they have asymmetrical karyotypes and aneuploid chromosome numbers with a range of 16 to 24 chromosomes and in general appearance of chromosomes. The species of *Halophila* are cytologically more related together. The species *of Halodule* also are cytologically related together.

Suitable conservation measures should be made to preserve these two rare monoecious species *(H. beccarii* and *H. decipiens)* which were collected only from Porto Novo and Pitchavaram areas. Balasubramanian *et al.* (2000) studied the antibacterial effect of the extracts of *H. ovata, H. ovalis* and *Halodule pinifolia* and the last species having more antibiotic effect. But studies were not made on the antibiotic effects of *H.becarrii* and *H. decipiens* occurring in the Vellar estuary at Parangipettai and Pitchavaram. All the seagrass species may be exploited for medicinal and other uses since they contain vitamins, iodine and various mineral substances.

Sex chromosome studies have been undertaken in dioecious species like *Halophila ovalis, H. ovata* and *H. ovalis* var. *ramamurthiana.* The 'XX' sex chromosomes of the female plants and 'XY' sex chromosomes of the male plants are shorter than the other somatic chromosomes of the respective species and they are acentric i.e. without constrictions. The Y chromosomes are shorter than the X chromosomes of the respective male plants in all the species studied. In *Najus minor* var. *spinosa* of Najadeceae also, the studies on sex chromosomes in male and female plants reveal some fact as the members of Hydrocharitaceae and Potamogetonaceae are present in the sea (Subramanian, 2000).

The characters of leaves, bracts, stipule, male and female flowers, habit and habitat of the five species and one variety of *Halophila* are more or less similar. Whatever differences found among the 5 species of *Halophila* are also found between *H. ovalis* and *H. ovata* var. *ramamurthiana.* So, this variety must have been already raised to species level. But, as far as morphological and cytological studies are concerned, all the species of *Halophila* studied herein may be considered as varieties of *Halophila ovalis.* Similarly all the species of *Halodule* may be varieties of *H. pinifolia.*

SYSTEMATIC TREATMENT KEY TO THE SEAGRASS FAMILIES

1a. Leaves ligulate male flowers without tepals, fruits indehiscent .... Potamogetonaceae

1b. Leaves eligulate, male flowers with 3 to 6 tepals, fruits dehiscent .... Hydrocharitaceae

I. Key to the genera of Hydrocharitaceae.

1. Leaves differentiated into petioles and blades, lamina oblong, elliptic, linear ovate, obovate or spathulate without tannin cel ls.. 2.

2a. Perianth in 2 - rows, male flowers numerous in spaths, pollen grains free peduncles long coiled in fruit .... *Enhalus*.

2b. Perianth in 1 row, male flowers single in spaths, pollen grains moniliform tube peduncles short, not coiled in fruit .... 3.

3a. Styles 2-3, fruits ca 5 by 3 mm across smooth, indehiscent .... *Halophila*

3b. Styles 6 to 8, fruits ca 20 by 18 mm across, echinate, dehiscent ... . *Thalasia*

II. Key to the genera of potamogetonaceae

1. Leaves terete, fleshy, grooved along the adaxial side, nerves absent, flowers in dichasial cymes .... *syringodium*.

2. Leaves flat not fleshy, without grooves, nerves present, flowers solitary or in pairs 3

3a Rhizones usually moniliform with scales, nerves 3, lamina 0.25 to 4 mm size 4

3b Rhizones not moniliform, without scales, nerves 7 – 22, lamina 4 to 10 mm size 5

4. Anthers attached at equal levels, with apical appendages, styles divided into 2 stigmas, fruit angular .... *cymodocea*

5. Anthers attached at unequal levels, without apical appendages, styles undivided, fruits globose .... *Halodule*

Eventhough *Thallasodendron* is represented to be present in terrestrial habitat, it is not so far available from Tamil nadu or in any part of India. Therefore the author is trying to find out from India seawaters.

MORPHOLOGICAL STUDIES

More important characters of the taxa are given below:

1. *Enhalus acoroides L.F. Royle (Plate – 1 Figs. 1 to 18)*

Dioecious species, plants are robust, to 2 m long with rhizome creeping, aerial stem stout about 2 cm thick. Aerial stem with basal decayed leaves, and upper 4 to 5 ribbon like linear with obtuse apex; male inflorescence is up to 6 cm long, immersed inside water, enclosing a large number of male flowers with 2 large lanceolate spaths. Female inflorescence is up to 6 to 7 feet length dilated at apex and subtended by 2 equal spaths. After fertilization, it contracted inside water. This plant is a giant modified form of Vallisneria spiralis a member of fresh water belonging to this family, Hydrocharitaceae, in having similar vegetative as well as reproductive characters and pollination mechanism. In this aspect of reproductive characters, Enhalus acoroides is totally different from those of Thalasia and Halophila species available in Coromandel Coast of Tamil nadu.

2. *Halophila beccarii Asch (Plate 6; Figs. 1 to 9)*

Plants monoecious, small and delicate plant when compared to Enhalus acoroides the larger and robust plants. There are 6 to 12 leaves at each node, leaves oblong lanceolate with distinct petiole and blade, male and female flowers simple borne in successive lateral branches or distantly. Each flower male or female is subtended by 2 to 3 spaths. The important point is that the leaf is having only a mid vein and cross veins are absent. It is available in sheltered localities of bays, estuaries, backwaters and in the undergrowth of mangrove swamps on sandy or clayish soil.

3. *Halophila ramamurthiana D.S. (Plate 3; Figs. 1 to 14)*

(*H. ovalis* (R. Br.) Hook f. subsp. Ramamurthiana Ravikumar and Ganesan). Plants dioecious, leaves 2.5 to 7.5 cm long, glabrous, 2 leaves at each node, lanceolate, mid veins and side veins clear, asymmetrical cuneate to attenuate at base, cross veins 7 – 16 paired petiole not sheathing. Seeds 6 – 12 in a fruit. It is more related to Halophila ovalis. This plant is rare and only available in backwaters of Marakkanam of Tamil Nadu and Theetapuram, Prakasam district of Andhra Pradesh.

This subspecies has been raised to the level of a species and named as *Halophila ramamurthiana* sp. nomina D.S. based on cytological and morphological studies.

4. *Halophila ovata* **(Plate 4; Figs. 1 to 11)**

Plants dioecious, plants small, spreading rhizome on the mud leaves 1.1 to 3.7 cm. long, paired at each node, glabrous, lamina mostly ovate, rarely oblong or oblong elliptic, cuneate or attenuate or rarely slightly oblique at base, mid vein and 3-9 side veins present.

It is available at pulicat lake of Chengalpattu district, Cuddalore O.T. Thanjavur, Ramanathapuram, Tuticorin, Pudukkottai and a few places of Andhra Pradesh.

5. *Halophila stipulecea (Forsk.) Asch*

Plants dioecious, leaves linear oblong, 2 leaves at each node, petiole sheathing at base, obtuse at apex, 1-3.5 cm by 1.5 – 5 mm size serrulate at margins, hairy on both surfaces, cross veins 4 to 14 pairs, intramarginal veins not merging with the margins. This species is restricted in distribution found only in Thanjavur, Pudukottai and Ramanathapuram district of Tamil Nadu. It is a marine form, not available in back waters.

6. *Halophila beccarii Asch (Plate 6; Figs 1 to 9)*

Plants monoecious, shoots erect, up to 2.4 cm long, rarely branched with a pseudowhorl of 6-11 leaves at each node, petiole sheathing, leaves oblong lanceolate, cuneate at base, acute at apex, 5-16 by 1-3 mm size, entire or minutely spinulose at margins, glabrouse, cross veins absent, intramarginal nerves 2, joining at apex with conspicuous mid-vein.

This plant is distributed continuously along the coromandal coast and is common in shallow, sheltered localities in bays, estuaries, backwaters and in the under growth of mangrove swamps, whereas absent in open sea. It is deeply seated in ground so washed ashore materials are rare.

It was collected in Cuddalore O.T. Portonova, Pichavaram of Cuddalore district, Tuticorin of V.O. Chidambaranar district and in a few places of Andhra Pradesh.

7. *Halophila decipiens Ostenf (Plate 5; Figs 1 to 12)*

Plant monoecious, branched, slender, a pair of leaves at each node; leaves oblong elliptic, cuneate at base, obtuse to acute at apex 10-25 by 3-6 mm size, serrulate along margins, hairy on both surfaces, cross veins 5-9 pairs, joining the intramarginal veins.

It is a rare plant, available in Alankarathattu and Thirespuram of Tuticorin V.O. Chidambaranar district of Tamil Nadu. It is unlike species of Halophila studied, not associated with any other seagrasses.

8. *Thalassia hemprichii (Ehrenb.) Asch (Plate 7; Figs 1 to 9)*

Rhizomes creeping, branched brittle, root single at each node leaves linear, falcate, slightly narrow at base, obtuse, rarely emarginated, or oblique and serrulate at apex; 15.5 by 1.2 cm size, nerves 7-13, parallel joining the intramarginal nerves at apex, flowers uniflorous, subtended by spaths. Peduncle of male inflorescence 3 cm long, female flowers subsessile. It is a rare marine form, available. It is available in Ramanathapuram district but washed ashore materials are available in Thanjavur and Pudukkottai districts.

9. *Cymodocea rotundata Ehrenb*

Plants dioecious, rhizome creeping branched. Shoots erect upto 37 cm tall with persistent, closed leaf scars 3-4, leaves per shoot, leaves linear, usually falcate, narrowed at base, obtuse, emarginate, oblique, rarely faintly serrulate at apex, 2.7 – 18.5 cm by 3-6 mm size, nerves 9 to 14, rarely reaching the apex, male flowers stalked; female flowers sessile.

This plant is a pure marine form available in all districts of Tamil Nadu, excepting Cuddalore and Kanyakumari Districts.

10. *Cymodocea serrulata (R.Br.) Asch (Plate 8; Figs 1 to 9)*

Rhizomes creeping, branched, flexible, shoots up to 70 cm long, erect, unbranched, bearing 2 to 5 leaves at each, open leaf scars, leaves linear often falcate narrowed at base, obtuse emarginated or oblique, and serrulate at apex, 23 cm x 6 = 10 mm size, nerves 12 – 22, midrib conspicuous, male flowers solitary; female flowers in pairs.

This plant is available throughout Tamil Nadu but absent in Andhra Pradesh. It is a pure marine form and not found in backwaters or estuaries.

11. *Halodule pinifolia (Miki) Hartog*

Plants dioecious; rhizome creeping, branched, often moniliferous; shoots up to 40 cm long, erect, branched or unbranched, bearing 2 to 3 leaves at each branch; leaves linear, narrowed at base, 10-29 cm long, 0.75 – 1.5 mm broad entire along margins, nerves 3 midrib prominent, lateral margins intramarginal, inconspicuous, leaf tips obtuse; male flowers 2, subequal, dorsally connate at the apex of the stalks; female flowers sessile enclosed by leaf sheaths.

This species has 2 forms like *H. uninervis*. This is a common species throughout Coromandal Coast, particularly in backwaters, estuaries, indirectly over swamps.

12. *Halodule uninervis (Forsk.) Asch (Plates 10; Figs 1 to 9)*

Plants dioceious, rhizomes creeping, branched, flexible, often monoliferous, shoots up to 30 cm long, erect, branched or unbranched bearing 2-4 leaves at each branch; leaves linear, narrowed at base, 6-24 cm by 0.7-5 mm entire margins, nerves 3 midrib conspicuous, furcated or widened at apex; male flowers 2, subequal; female flowers sessile, enclosed in leaf sheaths.

It is a common seagrass along the entire Coromandel Coast. It is found to occur in open seas, sheltered localities of bays, gulf, backwaters, estuaries and the margins of mangrove creeks.

13. *Halodule wrightii Asch (Plates 9; Figs 1 to 7)*

Plants dioecious; rhizomes creeping, branched, often moniliferous, flexible, shoots up to 15 cm long, erect, branched or unbranched, bearing 2 to 3 leaves at each branch; leaves linear, narrowed at base, 5 – 12 cm x 0.5 to 10 mm size, entire along margins nerves 3, midrib prominent leaf tip with 2 tooths; male flowers, subequal, female flowers sessile, enclosed by sheaths.

It is available at payhiar estuary, Akkamadam, Palk Bay, Rameswaran of Tamil Nadu and in a few places of Andhra Pradesh.

14. *Syringodium isoetifolium (Asch) Dandy (Plates 10; Figs 1 to 10)*

Dioecious; Rhizomes creeping, branched, slender, fleshy; shoots upto 60 cm long, erect branched or unbranched; bearing 2 to 3 leaves at each branch; leaves terete, narrowed at base, upto 40 cm long 1.5 mm thick, fleshy, brittle green; inflorescence upto 29 cm long, flowers in terminal cymes, the lower branches dischasial, upper branches monochasial, enclosed by reduced leaf sheaths and scales, male flowers upto 1.4 cm long; female flowers 5.9 mm long sessile.

The distribution of this plant is restricted to Palk Strait, Palk Bay, and Gulf of Mannar. It is not available in Andhra Pradesh. Recently, it was collected by Asin Selin Kumar et al., (2010) at Leepuram coast of Kanyakumari district of Tamil Nadu.

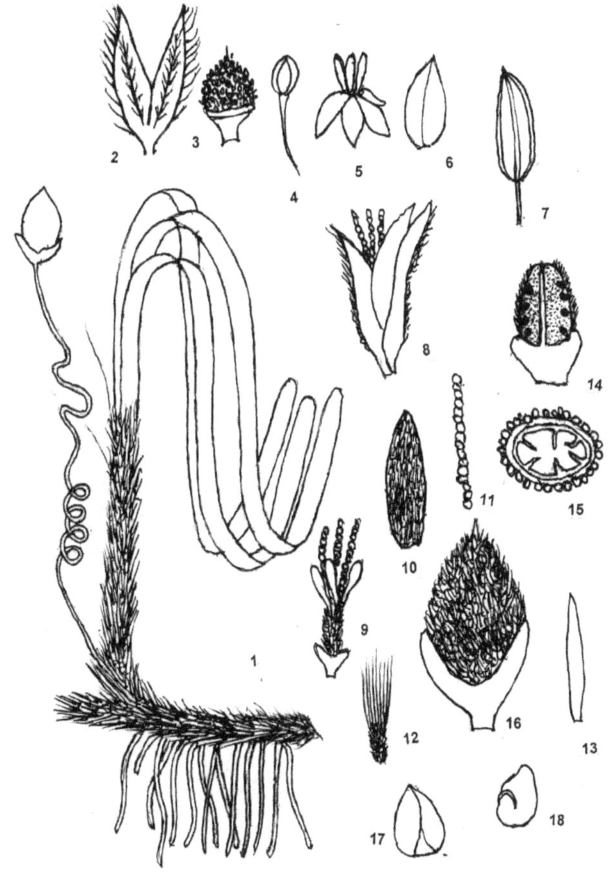

Plate - 1   *Enhalus acoroides* Fig. 1 = Plant; 2 = Bract; 3 = Carpel; 4 = Male flower bud; 5 = Male flower; 6 = Tepel; 7 = Stamen; 8 = Female flower; 9 = Stamens and carpels; 10 = Petal; 11 = Stigma; 12 = Calyx; 13 = Tepal; 14 & 15 = L.S. and T.S. Ovary; 16 = Fruit; 17 and 18 = Seeds.

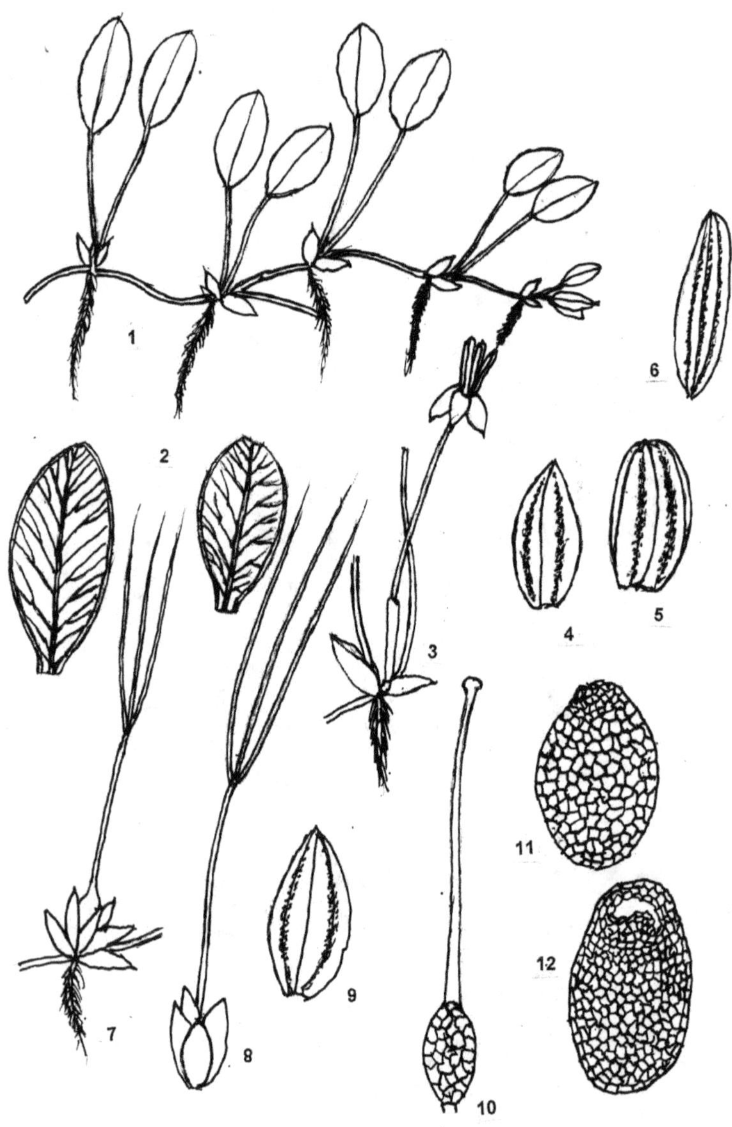

Plate - 2   *Halophila ovalis* Fig. 1 = Plant; 2 = Leaves; 3 = Male flower; 4 to 6 = Petals; 7 and 8 = Female flower; 9 = Petal; 10 = Carpel; 11 and 12 = Ovules.

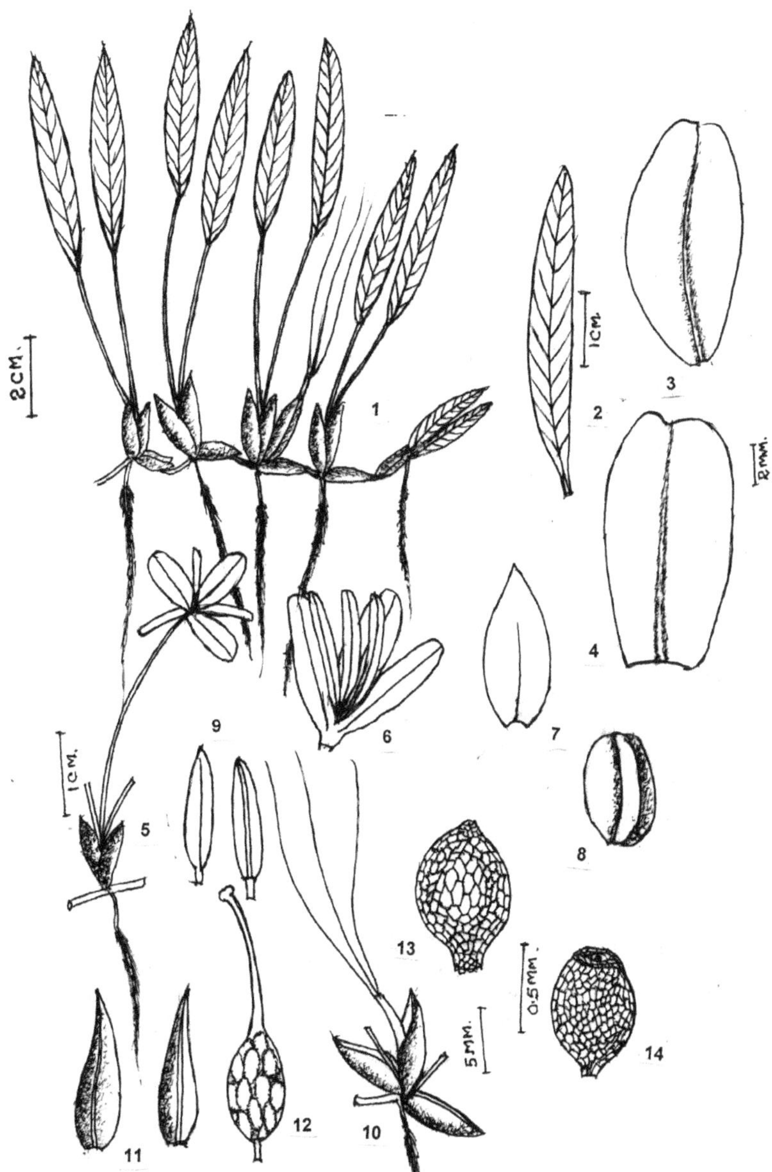

Plate - 3   Halophila ramamurthiana Fig. 1 = Plant; 2 = Leaf; 3, 4, 7 and 8 = Petals; 5 and 6 = Male flower; 9 = Stamens; 10 = Females flower; 11 = Petals; 12 = Carpel; 13 and 14 = Ovule.

Plate - 4   *Halophila ovata* Fig. 1 = Plant; 2 = Leaves; 3 and 4 = Petals; 5 = Male flower; 6 = Stamen; 7 = Female flower; 8 = Petals; 9 = Carpel; 10 and 11 = Ovules.

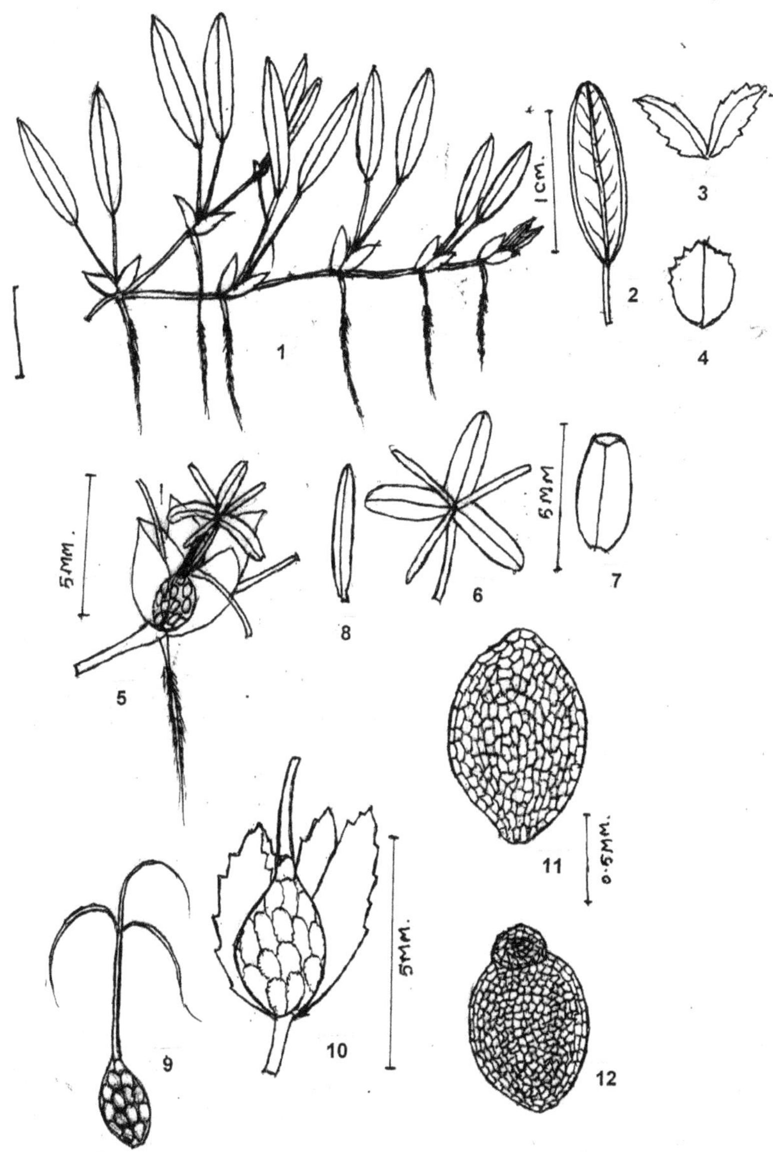

Plate - 5 *Halophila decipiens* Fig. 1 = Plant; 2 = Leaf; 3 = Bract; 4 = Bracteole; 5 = Male flower; 6 = Male flower; 7 = Petal; 8 = Stamen; 9 = Female flower; 10 = Ovary; 11 and 12 = Ovules.

Plate - 6   *Halophila beccarii* Fig. 1 = Plant; 2 = Leaf; 3 = Small leaf; 4 = Male flower; 5 = Stamen; 6 = Female flower; 7 = Ovary; 8 and 9 = Ovules.

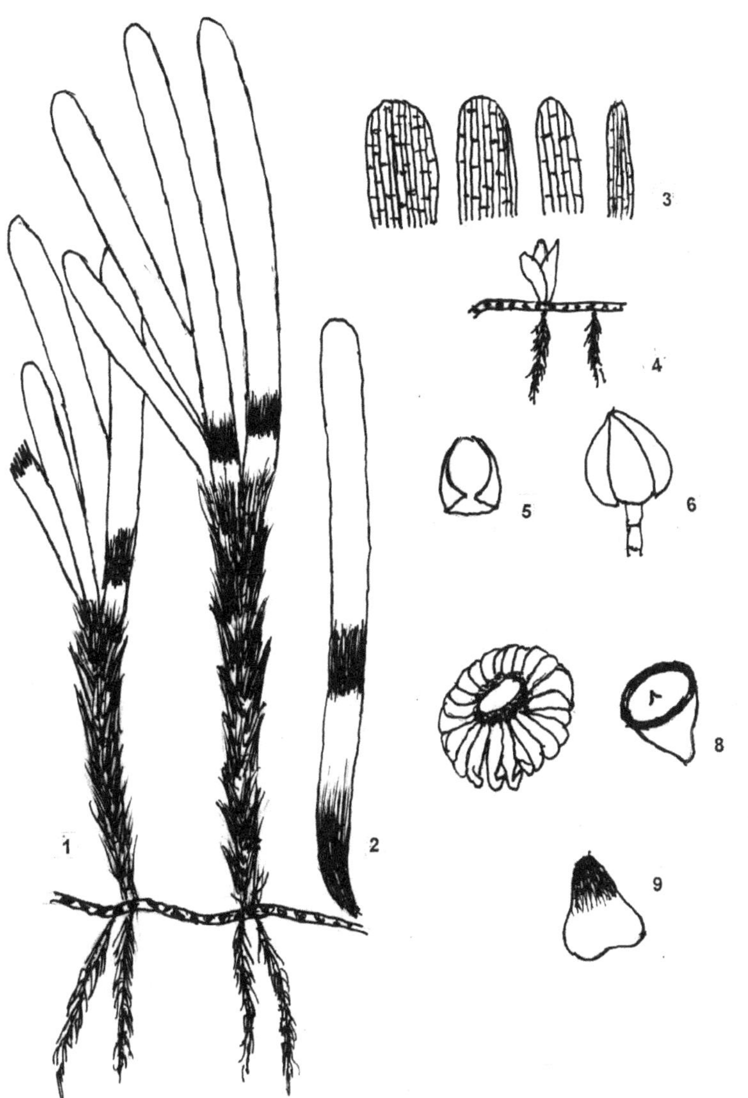

Plate - 7   *Thalassia hemprichii* Fig. 1 = Plant; 2 = A leaf; 3 = Leaf tips; 4 = Seedling; 5 = Scale; 6, 7 and 8 = Fruits; 9 = Seed.

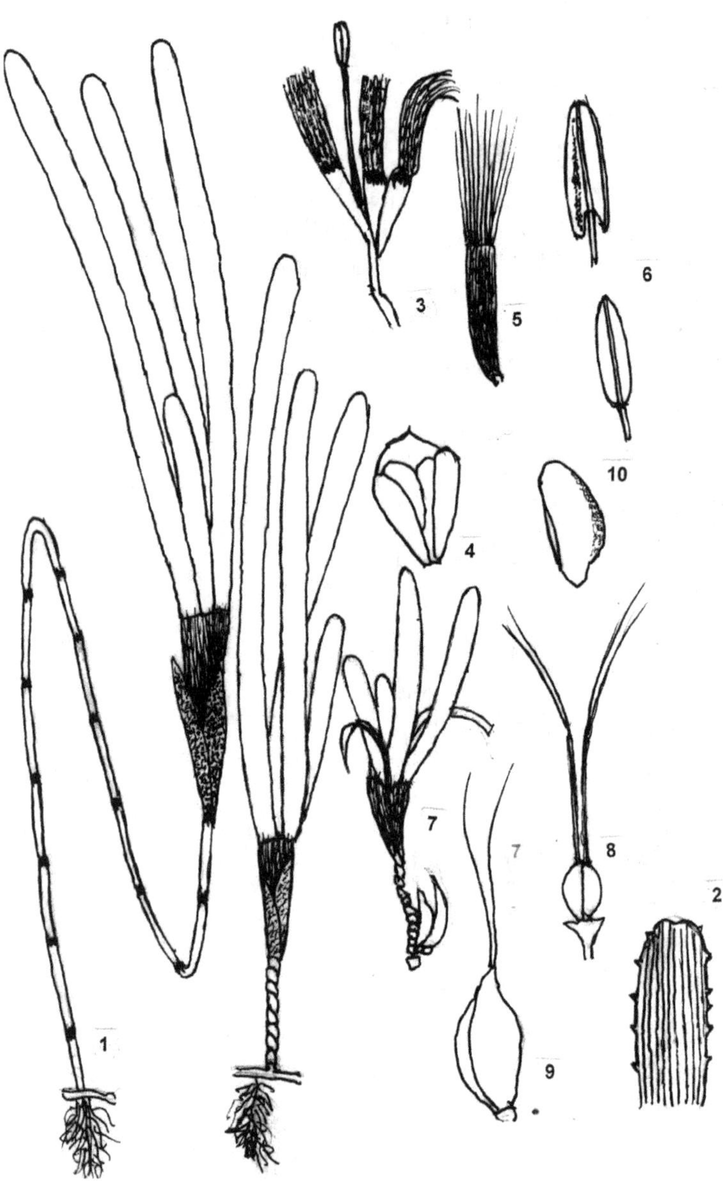

Plate - 8 Cymodocea serrulata Fig. 1 = Plant; 2 = Leaf tip; 3 = Male flower; Fruit; 5 = Petal; 6 = Stamen; 7 = Seedling with sex organs and carpel; 8 = Carpel; 10 = Seed.

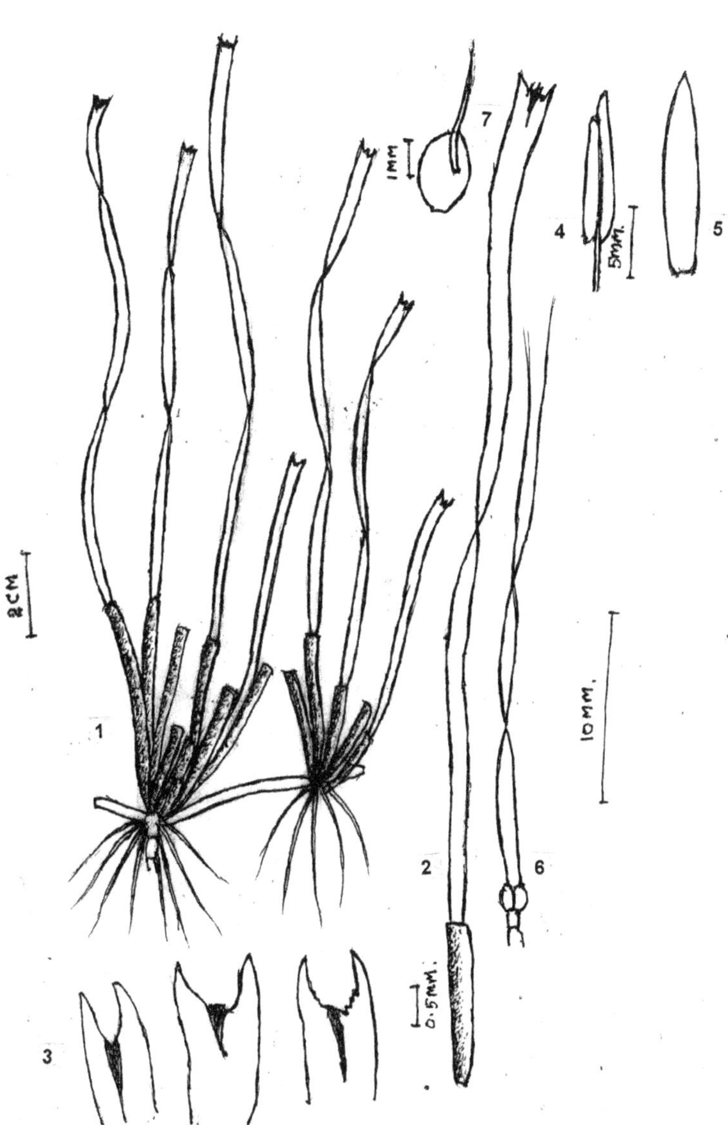

Plate - 9 *Halodule wrightii* Fig. 1 = Plant; 2 = Leaf; 3 = Leaf tips; 4 = Stamen; 5 = Petal; 6 = Carpel; 7 = Fruit.

Plate - 10   Halodule uninervis Fig. 1 = Plant; 2 = Leaf; 3 = Leaf blade; 4 = Leaf tips; 5 = Seed; 6 = Stamens; 7 = Carpel; 8 = A Female flower; 9 = Fruit.

Plate - 11 *Syringodium isoetifolium* Fig. 1 = Plant; 2 = A portion of plant; 3 = Male shoot; 4 = Male flower; 5 = Stamen; 6 = Female shoot; 7 = Female flower; 8 = Carpel; 9 = T.S. Ovary; 10 = Fruit.

REFERENCES

1. Arumugam, R., Ragupathy Rajakannan, R., Arivuselvan, N. and Anantharaman, P. (2010). Antimicrobial potential of some grasses against phytopathogens. *Seaweed Res. and Utiln.* 32 (1 & 2): 177-183.

2. Balasubramanian, R., L. Kannan and T. Thangaradjou. (2000). Screening of sea grasses for antibacterial activity against bacterial pathogens. *Seaweed Res. and Utiln.* 22 (1 & 2): 101-106.

3. Fedorov, A.N.A. (1974). *Chromosome number of flowering plants* Reprint by Koeltx Science Publishers, Koenictein, West Germany. 624 p.

4. Harada I. (1951). Karyotype and the development to chain like pollen of seagrass *Halophila ovata*. *Jpn J Genet.* 26: 226.

5. Ramamurthy, K., N.P. Balakrishnan, K. Ravikumar and R. Ganesan. (1992). Seagrasses of Coromandel Coast, India. *Bot Survey India*, Calcutta

6. Ravikumar K. and R. Ganesan. (1990). A new subspecies of *Halophila ovalis* (R.Br.) J.D. Hook (Hydrocharitaceae) from the eastern coast of Peninsular India. *Aquat. Botany*, 36: 351-358.

7. Ravikumar K., R. Ganesan and K. Ramamurthy. (1990). First report of a seagrass *Halodule wrightii* Asch (Potamogetonaceae) from *India. Journ. Econ. Tax. Bot.* 14(3): 711-714.

8. Stebbins G.I, J.A. Jenkins and M.S. Walters. (1953). Chromosome and phylogeny in Compositae, tribe Cichorieae, *Univ. Calif. Public. Bot.* 26: 401-430.

9. Subramanian D. (1988). Studies of Mangrove plants of Tamil nadu. *Cytologia.* 53(1): 87-92.

10. Subramanian D. (2000). Cytological studies in salt water species of angiosperms. *Seaweed. Res. and Utiln.* 22 (1 & 2): 159-160.

11. Subramanian D. (2010). Cytotaxonomical and cytogenetical studies on Marine Angiosperms of Coromandal coast, Tamil nadu. *Seaweed Res. And Utiln.* 32(1 & 2): 141-151.

12. Virendra Kumar and B. Subramaniam. (1982). Chromosome atlas of flowering plants of the Indian Subcontinent. Vol. II. *Bot. Surv. India.* Calcutta.

EXPLANATIONS OF PLATES AND FIGURES

Plate - 1: *Enhalus acoroides*

Fig. 1 = Plant; 2 = Bract; 3 = Carpel; 4 = Male flower bud; 5 = Male flower; 6 = Tepel; 7 = Stamen; 8 = Female flower; 9 = Stamens and carpels; 10 = Petal; 11 = Stigma; 12 = Calyx; 13 = Tepal; 14 & 15 = L.S. and T.S. Ovary; 16 = Fruit; 17 and 18 = Seeds.

Plate - 2: *Halophila ovalis*

Fig. 1 = Plant; 2 = Leaves; 3 = Male flower; 4 to 6 = Petals; 7 and 8 = Female flower; 9 = Petal; 10 = Carpel; 11 and 12 = Ovules.

Plate - 3: *Halophila ramamurthiana*

Fig. 1 = Plant; 2 = Leaf; 3, 4, 7 and 8 = Petals; 5 and 6 = Male flower; 9 = Stamens; 10 = Females flower; 11 = Petals; 12 = Carpel; 13 and 14 = Ovule.

Plate - 4: *Halophila ovata*

Fig. 1 = Plant; 2 = Leaves; 3 and 4 = Petals; 5 = Male flower; 6 = Stamen; 7 = Female flower; 8 = Petals; 9 = Carpel; 10 and 11 = Ovules.

Plate - 5: *Halophila decipiens*

Fig. 1 = Plant; 2 = Leaf; 3 = Bract; 4 = Bracteole; 5 = Male flower; 6 = Male flower; 7 = Petal; 8 = Stamen; 9 = Female flower; 10 = Ovary; 11 and 12 = Ovules.

Plate - 6: *Halophila beccarii*

Fig. 1 = Plant; 2 = Leaf; 3 = Small leaf; 4 = Male flower; 5 = Stamen; 6 = Female flower; 7 = Ovary; 8 and 9 = Ovules.

Plate - 7: *Thalassia hemprichii*

Fig. 1 = Plant; 2 = A leaf; 3 = Leaf tips; 4 = Seedling; 5 = Scale; 6, 7 and 8 = Fruits; 9 = Seed.

Plate - 8:   *Cymodocea serrulata*

Fig. 1 = Plant; 2 = Leaf tip; 3 = Male flower; Fruit; 5 = Petal; 6 = Stamen; 7 = Seedling with sex organs and carpel; 8 = Carpel; 10 = Seed.

Plate - 9:   *Halodule wrightii*

Fig. 1 = Plant; 2 = Leaf; 3 = Leaf tips; 4 = Stamen; 5 = Petal; 6 = Carpel; 7 = Fruit.

Plate - 10:   *Halodule uninervis*

Fig. 1 = Plant; 2 = Leaf; 3 = Leaf blade; 4 = Leaf tips; 5 = Seed; 6 = Stamens; 7 = Carpel; 8 = A Female flower; 9 = Fruit.

Plate - 11:   *Syringodium isoetifolium*

Fig. 1 = Plant; 2 = A portion of plant; 3 = Male shoot; 4 = Male flower; 5 = Stamen; 6 = Female shoot; 7 = Female flower; 8 = Carpel; 9 = T.S. Ovary; 10 = Fruit.